# Thinking about Animals in the Age of the Anthropocene

# Ecocritical Theory and Practice

Series Editor: Douglas A. Vakoch, California Institute of Integral Studies, USA

*Ecocritical Theory and Practice* highlights innovative scholarship at the interface of literary/cultural studies and the environment, seeking to foster an ongoing dialogue between academics and environmental activists.



**Titles in the Series**

*Thinking About Animals in the Age of the Anthropocene,* edited by Morten Tønnessen, Kristin Armstrong Oma, and Silver Rattasepp

*Romantic Ecocriticism: Origins and Legacies,* edited by Dewey W. Hall

*Linda Hogan and Contemporary Taiwanese Writers: An Ecocritical Study of Indigeneities and Environment,* by Peter I-min Huang

*Ecocritical Approaches to Literature in French,* edited by Douglas Boudreau and Marnie Sullivan

*The Green Thread: Dialogues with the Vegetal World,* edited by Patícica Vieira, Monica Gagliano, and John Ryan

# Thinking about Animals in the Age of the Anthropocene

Edited by Morten Tønnessen, Kristin Armstrong Oma, and Silver Rattasepp

LEXINGTON BOOKS
Lanham • Boulder • New York • London

Published by Lexington Books
An imprint of The Rowman & Littlefield Publishing Group, Inc.
4501 Forbes Boulevard, Suite 200, Lanham, Maryland 20706
www.rowman.com

Unit A, Whitacre Mews, 26-34 Stannary Street, London SE11 4AB



British Library Cataloguing in Publication Information Available

Names: Tønnesen, Morten, editor.
Title: Thinking about animals in the age of the Anthropocene / edited by Morten Tønnessen, Kristin Armstrong Oma, and Silver Rattasepp.
Description: Lanham : Lexington Books, 2016. | Includes bibliographical references and index.
Identifiers: LCCN 2016000941 (print) | LCCN 2016014449 (ebook) | ISBN 9781498527965 (cloth : alk. paper) | ISBN 9781498527972 (Electronic)
Subjects: LCSH: Animals (Philosophy) | Human-animal relationships.
Classification: LCC B105.A55 T45 2016 (print) | LCC B105.A55 (ebook) | DDC 590--dc23 LC record available at http://lccn.loc.gov/2016000941

∞™ The paper used in this publication meets the minimum requirements of American National Standard for Information Sciences Permanence of Paper for Printed Library Materials, ANSI/NISO Z39.48-1992.

Printed in the United States of America

# Contents

# Introduction

## *Once upon a Time in the Anthropocene*

*Morten Tønnessen and Kristin Armstrong Oma*

The stories we tell ourselves about our own species involve powerful narratives. Can they be trusted? *Once upon a time, in the Anthropocene . . .* we were human, we were powerful, and we were the talk of town amidst the chitter and chatter of the global animal community, from beast to bone. But what *is* the Anthropocene? No more than a word play, no more than just a sort of fairytale? Or *is* it, really, a proper designation for our current geological epoch here on Earth? Opinions differ, on this as well as on other aspects of the Anthropocene discourse.

The "Anthropocene" is a term that is rapidly gaining momentum. It presupposes that in recent Earth history humans took a leading role in driving changes on an unprecedented scale, with far-reaching consequences. The human species has spread to practically all habitable parts of the world, and the human population has increased tenfold since the industrial revolution and 1000-fold since the agricultural revolution. At the same time, the average real Gross Domestic Product per human being is currently more than ten times as high as it was around the time Columbus "discovered" the Americas some 500 years ago (Tønnessen 2008: 119–120).

Our current ecological footprint is linked to both what we consume, the amounts we consume, and how we consume it. A big portion of people on the planet consume more than is sustainable. This year, 2015, "Earth overshoot day" happened on August 13th, 6 days earlier than in 2014 (Withnall 2015). This means that the resources produced on Earth that year were used up by that day, and for the rest of the year the Earth is being over-taxed, with its land and resources depleted. This year, we will consume the equivalent of 1.6 Earths. "Earth overshoot day" is calculated by the NGO Global Footprint Network. Obviously, this affects creatures beyond the human species, including the other animals that share this planet with us.

There is no consensus on exactly when the Anthropocene began, as we will demonstrate in this introductory chapter. At any rate, in terms of geology the Anthropocene has been suggested as the name of our current *epoch*. A geological epoch is the shortest time interval referred to on a geological time-scale, with *periods* lasting for much longer (tens of millions of years), and *eras* lasting even longer (hundreds of millions of years). If formally accepted following the work of the Anthropocene Working Group of the International Commission on Stratigraphy, which is to present its recommendation in 2016, the term Anthropocene would take over from the term Holocene, which has until now designated the geological epoch since the most recent ice age (some 11.000 years ago). As Jan Zalasiewicz, the head of the Anthropocene Working group is quoted as saying in Kolbert 2010, "[w]e're trying to get some handle on the scale of contemporary change in its very largest context."

No matter what one thinks about the Anthropocene, the notion radically changes how we look at nature, and mankind. Ellis and Ramankutty 2013 suggest "anthropogenic biomes" as a suitable term representing "a contemporary view of the terrestrial biosphere in its human-altered form." They refer to research showing that "more than three quarters of Earth's land surface has been reshaped by human activity. Less than a quarter of Earth's ice-free land is wild," with only a fifth of this being forests. Here Ellis and Ramankutty apply the classical notion of wilderness as an area unaffected by humans. "Earth's remaining wildlands," write Ellis and Ramankutty, "account for only about 10 percent of global net primary production."

Ellis and Ramankutty's description of the state of the natural world is sobering:

> Humans are the ultimate ecosystem engineers, routinely reshaping ecosystem form and process using tools and technologies, such as fire, that are beyond the capacity of any other organism. This exceptional capacity for ecosystem engineering, expressed in the form of agriculture, forestry, industry and other activities, has helped to sustain unprecedented population growth, such that humans now consume about one third of all terrestrial net primary production, move more earth and produce more reactive nitrogen than all other terrestrial processes combined, and are causing global extinctions and changes in climate that are comparable to any observed in the natural record. Clearly, humans are now a force of nature rivalling climate and geology in shaping the terrestrial biosphere and its processes. As a result, the vegetation forms predicted by conventional biome systems are now rarely observed across large areas of Earth's land surface (Ellis and Ramankutty 2013).

The contributions to this volume are based on presentations at the conference "Animals in the Anthropocene: Human-animal relations in a changing semiosphere," which was held at the University of Stavanger, Norway, September 17–19, 2015.[1] This introductory chapter is focused on presenting the notion of the Anthropocene and how it changes the ways we think about nature, mankind, animals, and the relation between them.

New ideas are indeed needed. Almo Farina (2012, 18), one of the contributors to this book, claims that ecology has suffered "a sub-disciplinary fragmentation," and that it

> currently appears to be in difficulty when it comes to producing robust, key references with which to guarantee the sustainable development of human societies [. . .] This is largely due to the fact that ecology has assumed a distinction between human and natural processes that generates a continuous disaffection of man towards natural entities.

In other words, humans are seen as unconnected with the natural world at large. He calls for a different, more realistic epistemology of ecology, a cornerstone of which is to develop a notion of "resources" that is applicable in both ethology (the study of animal behavior) and economics (the study of human resource utilization).

The role and place of animals has so far received relatively little attention in the Anthropocene discourse, which has been dominated by references to climate change and other large-scale phenomena. But in the end, the living conditions of most if not all animals have to form an important part of what we regard as being at stake. Furthermore, on some accounts, animals have played a central role in how the Anthropocene emerged. As described in Tønnessen (2010, 98):

> By establishing a global *colonial organism* of sorts—to borrow a notion from biology—we have installed an ecological empire, hierarchically organised with *Homo sapiens* on top and with crop species, pets and livestock in privileged positions. Thereby we have further provided global breeding grounds for other species that might not otherwise have been able to spread at a global scale—from rats and doves to bugs and microbes of various sorts.

In short, it is our species' global reach, our status as "global species"—an anomaly in terms of usual biology—that has facilitated and caused the global reach that many animal and other species as well have today.[2] If it were not for the global human civilization, with settlements practically everywhere where land is habitable for our kind, the global makeup of animal ecology would have looked radically different. This is reflected in the fact that whereas around a tenth of global land surface was in year 1700 used for human food production, more than a third of terrestrial surface is today used for grazing or cropland (Bonnet and Woltjer 2008, 271).

The role and place of animals in the Anthropocene is definitively deserving of further academic studies. We hope that the near future will bring a degree of integration between the Anthropocene discourse as initiated and dominated by researchers in natural science, and Human-Animal Studies. While scholars within the humanities and social sciences in this context need to build on empirical studies from natural science, they can contribute, in the Anthropocene discourse, with a better understanding of the societal and cultural aspects of anthropogenic environmental change.

## WHICH OF US ARE "ANTHRO"?

The Anthropocene is named after the human species, using the Greek term "anthropos." The name implies that the members of the human species are both catalysts for the great changes to the sea, land, and atmosphere currently seen, and also the agents of this change. This book invites the reader to cast a wider net, and consider the role and place of animals in this geological age. But first, it is useful to probe into both the definition of "anthro" and, second, to consider the time depth of this age. These two questions are, as we will see, interlinked.

Donna Haraway finds the Anthropocene a difficult concept since it is by its very nature anthropocentric: "No species, not even our own arrogant one pretending to be good individuals in so-called Western modern scripts, acts alone; assemblages of organic species and of abiotic actors make history" (Haraway 2015, 159). From a feminist perspective, some (e.g., Rustick; Dow this volume) are contesting the concept of Anthropocene based on the tacit definition of which groups of humans are included in this definition of "anthro" (see also Lorimer 2015; Tsing 2015). According to a research report funded by the anti-poverty NGO Oxfam, 1 percent of the total population hold over half of the global wealth (Hardoon 2015, 2, see also Elliott 2015). Anecdotally, it might be claimed that a group of privileged, Western, upper class, middle-aged, white men are (or at least until quite recently have been) the policy-makers that have their hands on the rudder with regard to global issues including the environmental crisis. Looking at the gender, class, and ethnicity of world leaders and CEOs of global corporations, this is not far off the mark.[3] This critique surmises that "anthro" really only represents a minority of people on the planet, and that a privileged group of people hold most assets and power and are responsible for the lives of the many. The underlying rationale of this group seems to be a sense of entitlement and appropriation, which leads to actions that have far-reaching consequences beyond inequality and socio-cultural problems:

> The past twenty years have seen the proliferation of financial, administrative, and biological technologies for commodifying Nature—from ecosystem services to ecotourism to gene banks. Under the guise of naturalism or mastery, both of these approaches seek control over human and nonhuman life (Lorimer 2015, 5).

Among the consequences is consumerism that leads to the destructions of wildlife habitats as landscapes are torn up to make space for various production, extraction, or consumption facilities.

Those who argue against the designation Anthropocene on the basis of a deconstruction of a white, middle-aged, western male bias, aptly ask: What, then, of women, children and the elderly? And what of indigenous peoples that live lives separate from the Western world, deep in the rain forests of South America or the Gobi desert of Mongolia? Such questions have led to an increasing sense of discomfort with such a conflation of "anthro" with humankind (e.g., Rustick, this volume). This is a fair point, since the human species is heterogeneous, and one might argue that there is a Western bias inherent to the understanding of what it means to be human that underlies the Anthropocene. The Western understanding of the nature of human beings is of course not the only one. For example, Yanomami shaman Davi Kopenawa expresses deep concern over the consequences of destructive actions carried out by those he call "white men" in his native land in the rain forest, and says: "The white people, they do not dream as far as we do. They sleep a lot but only dream of themselves" (Kopenawa 2013, 313). Anthropologist Rane Willerslev lived with the hunters of the Yukaghirs and has portrayed how they, through both close observation and metaphysical communications (dreams and visions) sought to know, befriend, and engage with the moose they were hunting. Willerslev (2007, 2012) describes how the Yukaghirs saw themselves and the moose as living in a world of mirrors, in which both parties reflected aspects of each other (see also McNiven 2010).

Malm and Hornborg (2014) are among those who question "the use of the species category in the Anthropocene narrative" and argue "that it is analytically flawed, as well as inimical to action" (Ibid., 1). They suggest (Ibid., 1–2) "that the physical mixing of nature and society does not warrant the abandonment of their *analytical* distinction," and stress the continued relevance of social and cultural theory. Most discussion addressing the early as well as the current phase of the Anthropocene has focused on what Malm and Hornborg refer to as "the fossil economy," from our species' first mastering of fire to the currently massive burning of fossil fuels. Malm and Hornborg (Ibid., 1) remark that the fossil economy "was not created nor is it upheld by humankind in general," and stress that the modern fossil economy has developed in social settings with specific power relations, and so forth, that we must understand in order to understand the history of anthropogenic climate change. The two claim (Ibid., 3) that global economic "uneven distribution is a condition for *the very existence* of modern, fossil-fuel technology," given that "the affluence of high-tech modernity cannot possibly be universalized—become an asset of the species." Malm and Hornborg (Ibid.) therefore refute the "view of *humankind* as the new geological agent."

Nuances such as these are necessary, and Malm and Hornborg are right in stressing that social and cultural theory must supplement the picture painted by natural science. Still, to reject membership in "anthro" based on the assumption that only a small group of people are responsible for anthropogenic global change is in the end naïve and irresponsible. Such a perspective presupposes that the Anthropocene is understood as originating in quite recent history. Capitalism and the industrial revolution are fairly recent processes, with the industrial revolution taking place in the second half of the eighteenth century (the steam engine was invented in 1784). A proto-capitalist economy developed from around year 1500 and evolved into a capitalist economy around year 1820 (Maddison 2003). Only if this time scale is accepted as a starting point for the Anthropocene, claims (valid or not) could be made to the effect that many people do not partake in the systems, processes, or structures that have *caused* the Anthropocene. On the other hand, hardly any modern human being truly and consistently escapes consumerist society, regardless of their political or ethical inclinations, or indeed their gender.

Considering a greater time-depth to the Anthropocene results in a different perspective. An alternative understanding of the start of the Anthropocene is that it somehow originated with the agricultural revolution, which denotes the breakthrough of a new way of living that goes back some 10.000 years (cf. Mazoyer and Roudart 2006; Ruddiman 2003), with the domestication of the dog predating the agricultural revolution with tens of thousands of years (Bradshaw 2011). At this distant time, small groups of humans started the process of domesticating animals, plants, and landscapes, which is veritably a practice of the appropriation of nature. The mental change that underlies such an appropriation cannot be underestimated (Hodder 1990). For the first time, humans purposefully changed the life world and living conditions for other kinds of beings. Some of the first effective domestication of animals as well as plants took place in the Middle East and China some 9–10.000 years ago (Bonnet and Woltjer 2008, 263–4). In the following thousands of years, the practice of agriculture spread to large parts of the globe. And today, most of the inhabitants of the planet live according to a sense of rightful appropriation of our surrounding environment.

This aspect of the agricultural revolution provides time-depth that we can fruitfully utilize to gain an understanding of processes underlying the Anthropocene. The time period in which this shift happened is in archaeology referred to as the Neolithic revolution, and encompasses the domestication and modification of fauna, flora, and subsequently ecosystems (Barker 2006). Today, we all partake in the consumption of products that originated from the process of domestication. Thus, the Anthropocene is embodied, and has been building ours and other species' bodies for millennia. Contributing to this understanding is the fact that humans in different regions of the world have caused deforestation, soil salination, and erosion for several thousand years (Lines-Kelly 2004; on deforestation see also Ruddiman 2003). As a species, we have a long history of inflicting environmental change.

The speed of current changes, however, is unprecedented. For example, parts of the San Joaquin Valley in California are sinking nearly 2 inches every month due to ground water being pumped up and used for food production in the current drought (Farr et al., 2015). Here, again, the consequences of the Neolithic revolution are inextricably linked to sets of actions in which water that rained on the earth and has filtered through bedrock some 20.000 years ago is consumed in the blink of an eye.

We cannot escape the fact that it is humans, and living agents under domestication which are to varying extent in our control—in other words, humans and our recruited agents, so to speak (cf. the "global species" notion)—that affect the natural environment. Domestication is, from this perspective, a process that presupposes human mastery in the form of modification and manipulation of the behavioral dispositions, morphology, and life world of other beings, as well as of their distribution and whereabouts.

## DOMESTICATION

So let us consider the Neolithic revolution as a starting point for the Anthropocene. What happened at this particular period in time that led to such all-encompassing changes in the history of the planet? As of yet, this period rests in the mists of prehistory and no written sources can guide our understanding of driving factors.

When archaeology was a young discipline in the late 1800s, interpretations were guided by, and suffered the bias of, contemporary common sense-thinking. The Neolithic revolution was considered a shift based on purely economic reasons, and archaeology has played an important role in trying to explain these processes. Environmental archaeology has epistemologically been informed by a fundamentally humanist enlightenment paradigm. In a philosophical sense this entails that ontologically, the nature of being equals the nature of human being; the nature of action is of human action (Johannsen 2012, 305). Further, archaeology has traditionally been informed by a Cartesian worldview that considered humans as rational beings with souls whereas animals were seen to be like soulless machines driven by instinct, and bereft of reason. When combined, these perspectives presupposed that humans conquered animals, plants, and landscapes and subjected these to the human will, as though they were passive, even inanimate, matter. The supposed rationale underlying this appropriation was economic gain for the human masters. Even though archaeology is currently moving away from such deterministic models, the discourse is still being uncritically reproduced (for an overview of different explanations given for domestication see Russell 2012, 207–220).

The traditional discourse is the *domination discourse* in which domestic animals are regarded as beings controlled by humans for human consumption and other needs. This was succinctly expressed by anthropologist Tim Ingold in his article "From trust to domination," where he explained the transition from hunting to husbandry as a shift in attitude toward, and engagement with, animals from reciprocity to domination. Domestication implied that animals were kept as slaves (Ingold 2000; for critique see Armstrong Oma 2007, 2010; Knight 2005). His maxim was, by both archaeologists and academics from other disciplines, felt to express an underlying understanding of the shift in perception of animals that domestication caused, and has been widely quoted as an explanation of the relationship between humans and domestic animals as one of domination (for example Wengrow 2006, 81; Denham and White 2007, 9–11).

Several archaeologists claim that the process of domestication is much more complex (see Dransart 2002, 7; Jones and Richards 2003, 50; Whittle 2003, 80). Most notably, Steven Budiansky has advocated an alternative view and credits animals and plants with the domestication of humans (Budiansky 1992, 1994). Others have chosen a middle ground, and emphasize mutuality, fallibility, and chance in the development of humans and their environment (overview in Cassidy 2007; for example Zeuner 1963; Rindos 1984; O'Connor 1997). Terry O'Connor (1997) emphasizes human and animal co-evolution as a process that is beneficial to both, thus sliding toward an explanation influenced by symbiotic thinking. Rowley-Conwy and Layton (2011) characterize "[t]he modern farming environment or 'artificial steppe'" as "the ultimate form of niche construction by humans" (Ibid., 849), and add that hunter–gatherers in human history established ecological niches that were eventually used for agriculture. Farming, they claim (Ibid.), "originated not as a deliberate process of intensifying resource production, but as a series of small, accidental changes in the way that niches were constructed." For instance, systematic burning of vegetation changes vegetation patterns, the soil, and distribution of animals. However, Rowley-Conwy and Layton (2011, 856) also stress that the niche of animal domestication would not have been possible without the animal's social dispositions: "the social behaviour of certain wild ungulates, were what gave them the potential to be intensified to the point of domestication."

The anthropologist Natasha Fijn, who has studied Mongolian herders, points out that domestication of animals in terms of "making the animals into a being of human design, is likely to be a Western concept" (Fijn 2011, 18). She suggests that we should consider human–domestic animal relationships as "co-domestic," defined thus: "the social adaptation of animals in association with human beings by means of mutual cross-species interaction and social engagement" (Fijn 2011, 19). Dominique Lestel too thinks of human–animal relations, in a sense, as being on equal terms. "The ethologist has to be as creative as possible," he writes (2011,

98), "[t]he more creative he or she becomes, so the more complex and interesting the animal becomes." Without paying proper attention to the individual animal, behavioral traits including features of personality are not noticed. In Lestel's understanding, the capabilities of animals do "not allow for their objective [distanced, impersonal] description, because understanding them requires us to work *with* the animal (rather than *on* the animal)." Lestel (2002, 55) generally holds that "every human society is characterized by the nature of the hybrid communities its members establish with animals" and that these "hybrid communities are above all [. . .] semiotic communities."[4]

Thus, an understanding of the Anthropocene as initially established in the agricultural revolution must be rooted in an acknowledgment that it is not solely a human enterprise. Thinking of the agricultural revolution as a process of co-domestication gives us a different perspective on the Anthropocene, not one that is purely human. For example, Anna Tsing (2012) claims "science has inherited stories about human mastery from the great monotheistic religions. These stories fuel assumptions about human autonomy, and they direct questions to the human control of nature, on the one hand, or human impact on nature, on the other, rather than to species interdependence" (2012, 144). But, she posits, "What if we imagined a human nature that shifted historically together with varied webs of interspecies dependence?" If so, we could understand this age we live in as having been co-authored by humans and animals, and it is the deep interspecies entanglements that are a defining trait, rather than an understanding of humans as the sole doers and agents (Armstrong Oma forthcoming).

## THE ANTHROPOCENE—IN WHOSE INTEREST?

Some conservationists are concerned that the Anthropocene discourse has effects on our perspectives on nature and on conservation that are not at all constructive. The blurb of a recently published book by Edward O. Wilson (2016) includes this statement:

> Angrily challenging the fashionable theories of Anthropocenes, who contend that humans can survive alone in an Edenic bubble engineered for their own survival, Wilson documents that the biosphere does not belong to us. Yet, refusing to believe that our extinction is, as so many fear, predetermined, Wilson has written *Half-Earth* as a *cri de coeur*, proposing that the only solution to our impending "Sixth Extinction" is to increase the area of natural reserves to half the surface of the earth.

Here, as we see, Anthropocene "theories" are taken to imply the view that "humans can survive alone in [a] bubble engineered for their own survival." Of course, far from everyone involved in the Anthropocene discourse would support such a view—and many would outright reject it. Wilson, however, is not the only one who has expressed concern. "Although we agree that humans are a dominant species and have affected natural systems at a global scale," Caro et al. (2011, 1) write, referring to the Anthropocene discourse, "we suggest that humans may have less influence at smaller extents of specific regions and even ecosystems." They (Ibid., 1–2) "recognize that humans have had at least marginal influence on most if not all of the world's biomes," but state that "there are several reasons to doubt that humans have altered everything (a phrase that is generously interpreted as including nutrient flows and species composition and interactions)." Caro et al. (Ibid., 2) point out that the human footprint "shows large gaps at equatorial (central Africa), subtropical (central Australia, Sahara), temperate (Himalayas), and palearctic (Russia and Canada) latitudes," and that "increases in global temperature [. . .] have occurred principally in northern and southern latitudes and at high elevations." In the lower altitude tropics, where biodiversity is most abundant, temperature increases will be smaller than elsewhere.

For these reasons, Caro et al. (Ibid., 3) argue that there are several good reasons for acknowledging "that some areas of the globe are still intact." All wildernesses are not lost. In the context of the Anthropocene discourse, the authors (Ibid., 1) "fear that the concept of pervasive human-caused change may cultivate hopelessness in those dedicated to conservation and may even be an impetus for accelerated changes in land use motivated by profit." If no wilderness "is believed to be intact, it allows humans to think that species invasions are inevitable and not problematic and may open the floodgates to human manipulation of species assemblages" (Ibid., 3). "While accepting humans' enormous effect on the planet," then, Caro et al. (Ibid.) "see a crucial need to identify remaining intact ecosystems at local extents, to protect them, and to remind the public of them."

## WHEN EXACTLY DID WE BEGIN TO DOMINATE ECOSYSTEMS?

The Anthropocene notion was introduced in Crutzen and Stoermer (2000) to denote the "central role of mankind in geology and ecology." In the brief *Nature* article "Geology of Mankind," Crutzen (2002) claimed that the epoch of the Anthropocene "could be said to have started in the latter part of the eighteenth century, when analyses of air trapped in polar ice showed the beginning of growing global concentrations of carbon dioxide and methane." This coincides with the industrial revolution.

"One argument against the idea that a new human-dominated epoch has recently begun," writes Kolbert (2010), "is that humans have been changing the planet for a long time already, indeed practically since the start of the Holocene." This has led some, notably Ruddiman (2003), to defend an Early Anthropocene hypothesis. However, as Kolbert reports some of the members of the Anthropocene Working Group think, it could also be "argued that the Anthropocene has not yet arrived because human impacts on the planet are destined to be even greater 50 or a hundred years from now."[5]

Ruddiman (2003) claims that anthropogenic emissions of carbon dioxide ($CO_2$) and methane ($CH_4$) "first altered atmospheric concentrations thousands of years ago" (p. 261). What he refers to is "anthropogenic changes resulting from early agriculture in Eurasia, including the start of forest clearance by 8000 years ago and of rice irrigation by 5000 years ago" (Ibid.). In other words, he attributes the beginning of the Anthropocene to "the discovery of agriculture and subsequent technological innovations in the practice of farming" (Ibid.). Ruddiman claims that an observed increase in methane concentrations starting around 5000 years ago, which departs from an expected decrease, was caused by rice irrigation (Ibid., 262–265). A similar increase in carbon dioxide concentrations, which also departs from an expected decrease in concentrations (Ibid., 266), was, he claims, caused by pre-industrial land clearance related to establishment and expansion of agriculture (Ibid., 272-285). In his paper, Ruddiman focuses on both initial deforestation and re-cutting of forests following periods where plagues in effect led to temporary reforestation.

In an editorial comment to Ruddiman's paper, Crutzen and Steffen (2003), in support of their proposal that the timing of the beginning of the Anthropocene should be dated to the late eighteenth century, refer to the "astounding" consequences of the invention of the steam engine in 1784, along with medical advances and massive consumption of fossil fuels (p. 251). They also mention "[t]he Haber-Bosch industrial process to produce ammonia from $N_2$ in the air," claiming this "made the human population explosion possible" and remarking that "[i]t is amazing to note the importance of this single invention for the evolution of our planet" (Ibid., 252). Here as in other works (cf., e.g., Steffen, Grinevald, Crutzen, and McNeill 2011), Crutzen and Steffen also emphasize the great acceleration in environmental changes and societal developments that took place around 1950 (cf. also Steffen et al. 2004). "All components of the Earth System—atmosphere, land, ocean, coastal zone—are being significantly affected by human activities" (2003: 253). "The period of the Anthropocene since 1950 stands out as the one in which human activities rapidly changed from merely *influencing* the global environment *in some ways* to *dominating it in many ways*" (Ibid). In short, since 1950 or so several anthropogenic changes in the Earth System have been accelerating simultaneously, largely correlating with global economic growth.

In the end, however, Crutzen and Steffen (Ibid., 253) do acknowledge that there might be something to Ruddiman's Early Anthropocene hypothesis. "In summary," they write, "there may have been several distinct steps in the 'Anthropocene,' the first, relatively modest, step can have been identified by Ruddiman, followed by a further major step from the end of the eighteenth century to 1950 and [. . .] the very significant acceleration since 1950."

## WHAT NOW?

Just as there is no consensus on when the Anthropocene might have started, or whether it is accurate to hold the human species as such accountable for current environmental changes, nor is there any consensus on whether or not the Anthropocene is here to stay. Though few if any have claimed it explicitly, many might simply take for granted that from now on, the Earth will remain in the Anthropocene, with humans being decisive agents in global ecology. Will the Anthropocene ever end? This is in the end an empirical question. A moral version of the same question reads like this: *Should* the Anthropocene end, and if so perhaps as soon as practically possible? If so, is it a duty for morally and ecologically conscious human beings to work for a societal development that would eventually lead to an end of the Anthropocene? Some catchwords that can fruitfully be associated with an "out of the Anthropocene" stand include "de-growth" and James Lovelock's "sustainable retreat."

The notion of anthropogenic biomes, Ellis and Ramankutty argue, "moves us away from an outdated view of the world as "natural ecosystems with humans disturbing them" and toward a vision of "human systems with natural ecosystems embedded within them." Ellis and Ramankutty claim that this perspective "is critical for sustainable management of our biosphere in the twenty-first century." Their perspective here appears to take the Anthropocene as a given, rather than to challenge its right to be, at least in the quite near future.

As already mentioned, Crutzen and Steffen (2003) identify three distinct historical phases of the Anthropocene to date. They further envision "the fourth phase of the 'anthropocene,' which should be developed during this century" and should "not be further characterized by continued human plundering of Earth's resources and dumping of excessive amounts of anthropogenic waste products" (Ibid., 254–256). Instead, this fourth phase should be characterized by "vastly improved technology and environmental management, wise use of Earth's remaining resources, control of human and of domestic animal population, and overall careful treatment and restoration of the environment—in short, responsible stewardship of the Earth System."

Crutzen and Steffen's vision of a more environmentally friendly phase of the Anthropocene has progressive elements, but it leaves one important question unanswered: Namely, *should* the Anthropocene end, and if so, how soon *could* it end (assuming that both human and non-human quality of life will be safeguarded in the process)? Responsible stewardship is of course preferable to ecological developments to date—but are we morally justified in remaining "in power" in ecosystems worldwide? If one's answer to that question is "no," one would at the very least have to acknowledge that implementing an end to the Anthropocene would take a few centuries, since it would have to involve a carefully planned downsizing of the scale of human activities (and of the global human population).[6]

## NOTES

1. Minding Animals Norway was a co-organiser, along with the research project "Animals in Changing Environments: Cultural Mediation and Semiotic Analysis" (EEA Norway Grants/Norway Financial Mechanism 2009–2014 under project contract no. EMP151).

2. As detailed in footnote 66 in Tønnessen 2010 (p. 110), "Nobel laureate Paul Crutzen has read a version of "the article and "acknowledges that '[t]here is really a connection between our articles and with the concept of "Anthropocene" and your "global species"' (private correspondence)."

3. This said, it should also be noted that the rise of "the rest" the last 10-20 years has challenged the West's dominating role in the world economy. For instance, China is on path to overtake the role as the world's single largest economy in a few years, and has already become the world's biggest polluter, consumer of steel, and biggest car market, to mention but a few facts. Generally, lower- and middle-income countries are catching up with the West in economic terms.

4. As described in Tønnessen (2015, 15–18), comparative mapping of the *Umwelten* (lifeworlds) of various animals and their perception of humans and human artifacts can contribute to cultural theory as well as result in precise portraits of specific human communities.

5. Crutzen and Steffen (2003) write that sea level rise due to anthropogenic climate change caused by emissions until today and the rest of the twenty-first century is expected to amount to "0,5-10 m until the end of this millennium." The fact that the sea level might continue to rise for centuries even if we stop emissions in a few decades is often overlooked in the discourse on climate change.

6. For a review of various scenarios for future economic growth and world population until year 2300, see Tønnessen 2008.

# *I*

# Beyond Human Eyes

# ONE

# Held Hostage by the Anthropocene

*Susan M. Rustick*

What will my canine companions think if the Working Group on the Anthropocene makes an initial proposal that our current epoch be called the Anthropocene? What will the elm tree sense or the aronia bushes? What clarion call or trumpet of death will be heard by the Whooping Crane or the deer that sleep in prairie? Should the designation of a geological epoch acknowledge only humans and not all life on Earth? I argue that the proposed name "Anthropocene" is comparable to the experience of Narcissus, peering at his own human reflection in the pond. The pond itself does not exist for Narcissus; the trees and sky above the pond do not exist. All that exists is the gazer enrapt by his own image. The gazer remains in his reflexive universe, captivated by his mind's eye.

Is this an appropriate analogy for the current proposal to name an infinitesimal fraction of a second in the history of the Earth, the Age of Humankind? Why not name the current period in Earth's history after ourselves? What's in a name? The possibility of an epoch named the Anthropocene has taken off in popular culture and with some well-intentioned environmentalists (Monaskersky 2015). The name appeals either to a feeling of god-like omnipotence or to a recognition of what incredible damage we have done to the Earth and the biosphere in a blip of time. What's in a name? The problem is precisely this: that pop culture and a select group of humans designate the human as the most important feature of our current time. With the use of this term, humans go beyond the usually assumed position of "subject" to also designate the very same concept of human as the object of an epoch lasting, as epochs do, thousands of years. We, enrapt by the vision of ourselves in the reflecting pool of our mind, like Narcissus, are spellbound and held hostage. It is a closed, circular loop of self-fascination and self-entrapment, which demotes all other life and Gaia to mere background scenery. The name Anthropocene implies that if life on Earth is to be saved, we will be the ones to save it. If life on Earth is doomed, we are the ones who did it. In truth, our designation of this epoch holds us hostage in our exclusionary, dualistic perspective. Situated thus, we are unable to generate an appropriate ethical response grounded in compassion and a recognition of our mutual interdependence with all of life. This self-aggrandizing designation excludes all other life and is unjustly applied by the few designators to include all humans.

## THE DUALISTIC, DESIGNATING MIND

"Designating mind" is a Buddhist concept used to identify a step in the processes of consciousness (Hirschboeck 2015). It allows us to share language and symbols, to understand and respond to each other: to agree that a stop sign is red, has eight sides, and designates the meaning of the word "stop." On the other hand, the designating mind separates the "self" from the "other." We are the subject, separate from what we designate, the object. This dualistic thinking is a hallmark of Western philosophy and cultural norms. Non-dual realization of oneness sees dualistic thinking as "splitting the body of reality" (Martin 2014). According to Buddhism and most other non-Western worldviews, this failure to recognize all beings as part of one body of reality brings suffering. Because dualistic thinking keeps us from seeing how all the diverse beings are interdependent participants co-creating the whole, we are held hostage by our small egoic view of the world. Freedom from this trap requires a paradigm shift beyond the ego, an openness to multiple reference points beyond the human. Otherwise, the designating mind only appropriates the way of seeing, hearing, feeling, touching, tasting, smelling, and thinking which reinforce the self that has designated something in its own terms. "Appropriating mind" is at work when the self automatically lets some perceptions into consciousness and excludes others. For example, if I have been conditioned to designate chickens as food, I will most likely —and unconsciously—exclude perceptions of the bird as an intelligent, caring being. I will appropriate the smell of cooking chicken, not with revulsion, but with a conditioning that references memories of a pleasant taste. When the mind awakens to non-dual oneness, fixed designations and appropriations fall away. Our ethical response to the pain of the crises of the Earth is freed from the self-referential prison of our own minds. We are then capable of living as wise, compassionate individuals keenly aware of our interbeingness with all and making ethical choices accordingly.

## ALTERNATIVES TO DUALISTIC THINKING

This human-nature dualism is not a universally accepted view. Besides Buddhism, indigenous peoples' spiritualities, Hinduism, Jainism, and deep ecologists regard humans as embedded in nature and reject dualism as a wholly inaccurate view.

What would a First Nations' person call the current geological epoch? How would "The Age of Man" resonate with this worldview that sees all life as relatives, all life as agentive and equally important, to be regarded with mutual respect and without dualism or hierarchy? What kind of epoch would be named for human beings to the exclusion of all other sentient beings and even to the subordination of the Earth itself on which all life depends?

Chief Luther Standing Bear said that the Lakota revered nature and felt best when they were immersed in nature. The Lakota loved to sit on the ground, to live on the ground, and the elderly liked to walk barefoot on the Earth to feel its warmth and power. Kinship with other creatures was central to spiritual tradition and daily life. The Lakota Chief said, "kinship with all creatures of the earth, sky and water was a real and active principle. For the animal and bird world there existed a brotherly feeling that kept the Lakotas safe among them and so close did some of the Lakotas come to their feathered and furred friends that in true brotherhood they spoke a common tongue" (Gottlieb 2004, 39). To say that human beings and animal beings "spoke a common language" is both telling and compelling. Nature was not mute, mysterious, or a malicious force separated from humans, but a family to which one belonged and with whom humans shared a semiotic system. This relationship was founded on respect, gratitude, and humility. It was also founded on an ability to *listen* to other sign systems besides that of humans, and to make meaning in that encompassing semiosphere.

Jainism is an ancient religion that arose and flourished in India before spreading to adherents around the world. Followers in India number some 4.3 million, although most believe that number is deceptively low since many Jains refer to themselves as Hindu. A central tenet of Jainism is *ahimsa*, or nonviolence toward all sentient beings. The Indic karma of Jainism and Buddhism is not to be confused with a fatalistic world-view as the term is used in Western pop culture. Karma means both action and the law of cause and effect in that action which at one point affects all other parts. It is often imaged as a web wherein all life is interconnected, including human life, and all life has intrinsic worth regardless of the judgments conferred by our anthropocentric viewpoint. Human life is equal in value to any life form because an action by any individual or species in the threads of the web will eventually have an effect on everything else, including all other species, life forms, and other humans. While some in the West see power and dominance as superiority, Jains see humility and compromise as a sign of inner strength. Mahavira, the twenty-fourth Path-Finder of Jainism, said, "One who neglects or disregards the existence of earth, air, fire, water, and vegetation disregards his own existence which is entwined with them" (Rankin 2009, np). Rankin states that our greatest karmic delusion "is the delusion that we are separate from and above the rest of nature and have unlimited powers over it" (Ibid.). He points out that it is the very scale of the impact of humans on the rest of nature that reveals the folly of maintaining a philosophy of human-nature dualism. From this perspective, to choose the name Anthropocene only confirms and perpetuates both delusion and folly. The name implies that only other species are going extinct while in reality our actions are making it more likely that humans will go extinct also. This could be called an "unintended consequence," but it should not surprise us if one recognizes that everything we do vibrates over the web of life in which we are embedded. Our actions have consequences and "hence we are really not individuals but indivisibles" in a world grounded in mutuality and entailing responsibility (Rankin 2009). John Muir said, "When we try to pick out anything by itself, we find it hitched to everything else in the universe" (1911 [1988], 110).

Mohandas K. Gandhi was born a Hindu and remained a devout Hindu, but he was also greatly influenced by Jainism. Gandhi said, "As human beings our greatness lies not so much in being able to remake the world—that is the myth of the atomic age—as in being able to remake ourselves." Thus it is more urgent that we humans change our consciousness and value systems than our technology and politics. Transforming the world in terms of ourselves is not our greatness. It seems that it should be easier to remake one's self than to remake the world, but this is not possible when humans are of the mindset that humans are the pinnacle of creation, are separate from nature, and can fix whatever needs fixing. The name Anthropocene carries all these assumptions.

Vandana Shiva, a Hindu, is an environmental activist, a supporter of biodiversity, and anti-globalization author and teacher. She is the founder of Earth University in India, which "teaches Earth Democracy." Earth Democracy "is the freedom for all species to evolve within the web of life, and the freedom and responsibility of humans, as members of the Earth family, to recognize, protect, and respect the rights of other species. Earth Democracy is a shift from anthropocentrism to ecocentrism" (Shiva 2013, 50). Shiva calls the human-nature dualism an "eco-apartheid based on the illusion of separateness of humans from nature in our minds and lives" (2013, 49). This eco-apartheid is facilitated by the concept of "Dead Earth" as opposed to Mother Earth. These designations were catalysts for a philosophy touted by Descartes, by modern science, and then realized in the Industrial Revolution. Earth was commoditized and as a dead thing, a "resource" for human use, could be harmed with impunity.

## THE PERILS OF ANTHROPOCENIC THINKING

Carolyn Merchant, in her book *The Death of Nature,* argues that before the scientific revolution, the Earth was conceived of and named a living mother, and nature herself was alive. Since the Enlightenment, Earth and nature have been atomized and objectified, in effect treated as dead matter to be used for human purposes without compunction. This death also resulted in the end of a shared semiosphere with animals and humans. When mutual communication was lost, humans also became ignorant of the interspecies community which is ultimately the basis of the human sense of self.

Vandana Shiva's colleague Satish Kumar, editor of *Resurgence and the Ecologist,* advocates reverence for nature and a humble recognition of our species as just one of a multitude in the community of life on Earth. He writes that we would be wiser to call the new epoch the Ecozoic, a proposal first made by the late eco-theologian Thomas Berry. Berry "urged humanity to repair the damage it has inflicted on the Earth and to bring about an era that is respectful of Nature, self-renewing and ecologically sustainable. He envisioned a new age in which humans and all other species live in harmony with one another" (Kumar 2013, 1).

In the *Earth Island Journal* of spring 2013, Ginger Strand writes:

> [The Anthropocene] plays too slickly into the hands of the techno-utopians who will argue that since we're at the helm, we might as well put our hands on the rudder and steer. The very word "Anthropocene" makes too little accommodation for anything else besides us. It's not going to help us live with more grace in a world full of things we can't control, things we don't know, things we might never know.

Strand recognizes the self-congratulatory nature of the proposed term. Humans can feel they have a legitimate right to retain control and perhaps even do a bit more active management, such as with geo-engineering.

In the same edition of *Earth Island Journal,* which focuses on the proposed Anthropocene epoch, Maureen Nandini Mitra (2013) points out how the sounds of humans are creating such a din that we are drowning out the ability of other species to communicate; we are disabling the other voices of the semiosphere to a devastating degree. In her article "Extremely Loud: We have Drowned out the Natural Landscape," Mitra writes:

> Biologists have found that some birds in urban areas are finding it hard to hear each other and their young, which impairs chicks' growth, as they are less likely to be fed, leading to a decline in their numbers. In forests and deserts and plains a range of animals from gleaning bats to frogs to the endangered pronghorns in Arizona's Sonoran desert are abandoning their habitats in order to escape the noise of chainsaws and

> low-flying jets. The situation might be even worse under water. Ocean noise has been increasing by about three decibels every decade in the past 50 years due to sonar blasts by navies, shots from air guns used in deep-sea oil and gas exploration, and the thrum of cruise and freight ships. The cacophony disorients and sometimes leads to the death of marine animals, especially whales and dolphins that rely on their acute and highly specialized hearing for communication, navigation, and detecting predators.

As humans get louder and louder, we are creating a deafening noise that is impairing communication within species, between species, and within the whole of the semiosphere. This is not just an inability or unwillingness on the part of many humans to listen to the rest of nature, but an anthropogenic cause of other species becoming unable to listen to each other. We have put other animal cultures (de Waal 2001) in danger by making their communication systems dysfunctional. This indicates that the Anthropocene would not only be the age of humans, but it is also one of our making. Interestingly, we already have a label for the latter.

The Holocene was officially so designated only seven years ago, and it is an epoch that stretches back 11,000 years. But it has also been known as "The Anthropogene." There is only one letter difference between Anthropocene and Anthropogene, but the distinction in meaning and what it indicates about our current mindset is no small matter. An Anthropogene age is one which has been vastly impacted by human activity and denotes human agency. The Anthropocene, on the other hand, makes no reference to the nature of humankind's role. It simply designates the period as ours and supposedly ours for at least thousands of years hence. It is notable that only a couple of years after the last official designation, we are already considering a new term and a new period, given that an epoch is not a century or two or even a millennium. In fact, Sir Martin Rees, who is the chair of the Royal Society of London for Improving Natural Knowledge, states that given a best-case scenario of humans making sweeping, positive policy changes, current civilization has only a 50-50 chance of surviving another century (Oelschlaeger 2014, 232).

The act of naming someone or something is potentially an act of great consequence. It can destroy, denigrate, make invisible, exalt, restrict, or exclude. Naming, or designating, converts something like the biosphere and semiosphere, which are complex, multidimensional, and interconnected, into a two-dimensional entity, a caricature isolated from the unity of experience and reduced to a cliché. For example, a dog called Fluffy will evoke certain expectations which are different from a dog called Max. The name will in turn create a frame for perceiving the dog and impact the nature of interacting or not interacting with her. It might also conjure up an attitudinal perception toward the person who would name a dog "Fluffy," which conjuring would be embedded in the web of experiences and meaning making systems of the conjurer.

Halliday (2001) notes that our semiotic system is riddled with the concepts of human-nature dualism. Only humans can be a "he" or "she," while other animal beings may be called "it." This makes nonhumans less than humans, being objectified, stripped of gender, and put into the class of "things," implying lack of sentience and intelligence. European languages also demote our natural resources, such as air, water, soil, and fuel, by classifying them as noncount nouns. The only measures made of these are through anthropocentric units tied to their utility. This dualistic distinction is the kind of assumption that divorces humans from all other life, and creates a conception of Earth as context instead of source. Instead, ecofeminist Heather Eaton (2011) calls for a recognition of our interconnectedness and our "radical dependency" with all other life. Such recognition is absent when a geologic epoch is named only in human terms.

Wilhelm Trampe (2001, 232) reveals how language from agriculture also reinforces our objectification of the other-than-human and hence the illusion of a nature-human schism. He says that we reify other beings: "Living Beings are treated in accordance with the economic-technological ideology like objects that are produced, managed, optimized, and utilized." An example is the distancing language used in livestock reproduction, where a cow's embryo is called "recipient material." Euphemisms also promote dualistic thinking. Slaughtering an animal is called "harvesting" it. Poisoning plants is done with an herbicide and poisoning animals is done with a pesticide. The images evoked by pesticide and herbicide are that of something tidy, clean, and uniform, like a nice expanse of uniform lawn. Industrial agriculture is a respected power in the economy and society, and such operations are seen as efficient, productive, and profitable. The disastrous consequences for the sentient beings used in the system are not "visible" and neither are the consequences of the prolific use of toxic chemicals in the environment. On the other hand, a traditional, small farmer who might have an integrated, interdependent sense of her or his relationship with the plants and animal beings does not enjoy high status, but is rather regarded as a rustic, backward peasant.

## CHIEF GRIZZLY BEAR AND *GRIZZLY MAN*

The semiotic power of those in the position to designate is clearly exemplified by the European American immigrants' system of signs for the indigenous First Peoples of North America. One group of humans become the subjects of the European invasion, and the other group of humans, the original inhabitants, is objectified. As in all aggressions, the aggressors used language to designate the First Peoples as inferior, not really human, and especially so as the First Peoples are seen as part of nature in the culture-nature divide. Hence Anthropocene excludes indigenous people in its understanding of humankind.

An example from the White/Indian discourse helps to demonstrate the dysfunction when one semiotic system assumes hegemony over others. In the archives of the Wisconsin Historical Society, there is mention in an article, referring to 1830, of a Chief Grizzly Bear who went to the nation's capital for the purpose of promoting friendship, the assurance of a bit of land on which to live, and a small annuity (Ellis 1856). The President listened to Chief Grizzly Bear and his cohorts and then assigned the Secretary of War, General Eaton, and Colonel Stambaugh to be the commissioners to decide on the delegation's offer to sell their land. This situation is most interesting in its demonstration of contradicting constructions of meaning. It comes in a footnote. A reporter for a Maryland newspaper, the *Baltimore Republican,* noted that the Chief's name was not Grizzly Bear, but actually Great Speaker, "a person of great personal dignity" (Ibid., 434). The reason given was that it was more convenient to call the chief Grizzly Bear. This presents the question as to why would it be more convenient to call the chief Grizzly Bear? It was "convenient" for the European Americans because it reduced the Chief's status to a caricature of a leader of his people, it closely aligned the Chief with nature as opposed to culture, and it further usurped the power to call him whatever they wished, in effect insulting and denigrating the Chief as not a "Great Speaker" who has command of language.

A further semiotic lesson in this footnote occurs when the Chief's handlers take him on a tour of the Capitol. In the rotunda, Chief Great Speaker took it upon himself to affix his gaze to murals in the four directions, east, north, west, and south. To the east was a mural of the Pilgrims. Of this, Chief Great Speaker said, "There Injun give white man corn." To the north, there was a depiction of Penn's treaty, and the Chief said, "There Injun give 'um land." To the west where there was a painting of Pocahontas rescuing Captain John Smith, he said, "There Injun give 'um life." And then there was the *coup de grâce*: to the south was a painting of Daniel Boone with a knife in the chest of a Native American and his foot on another dead Indian. The Chief said "There white man kill Injun." The glorious history of the United States, as proudly displayed in the four directions of the Capitol rotunda, told a very different story, a story of goodwill and generosity repaid with treachery and mass extermination. At least the presence of the footnote indicates that this was a story that the White correspondent found impressive enough to find its way into a newspaper. The footnote states that "the old chief's conclusions and sarcasm are expressed with sententious brevity and striking effect" (Ibid.). This recounting of the situation hardly shows us an inferior human being, but rather one who can turn the presumptuous White man's semiosis of the murals on its head with a few epigrammatic interpretations.

This stands in stark contrast to Western views of nature, which have parallels with man's view of "women," another binary distinction that is entrenched in the designation of the other, and another who is "lesser than" and to be controlled. In the documentary *Grizzly Man,* (2005), Werner Herzog portrays Timothy Treadwell as a delusional, mentally unbalanced human who thought he could live with grizzly bears on an Alaskan nature reserve. The intended message seems to be that Treadwell is naïve, childlike, and is using nature as an escape fantasy and a way of avoiding the reality of a life a human should lead in human society. Herzog's voice in the film makes clear his view of nature and wilderness as dark, sinister, and ruthless. He mocks Treadwell and edits Treadwell's own posthumous film footage to portray a man coming unhinged, who mistakenly thinks he can communicate with bears. Alternate editing, done as a series on *Animal Planet* called *Grizzly Man Diaries,* presents a remarkably together man who actually does communicate well with the bears. The latter, however, was completely overshadowed by the popular documentary with its warning of the human-nature divide that must not be crossed. The fact that Treadwell did spend ten long summers with the grizzlies before he came to harm is thus deemed not noteworthy.

Even post-Christian, Western culture has, during certain periods, had a different relationship with nature. Consider the Romantic Period. Kalevi Kull (1998) notes that during this time the idea flourished that "nature speaks" to humans. However, the image of nature in Romanticism was of the same idealized, kind, and beautiful vision that Herzog ascribed to Treadwell as a sign of Treadwell's suicidal ignorance. Kull asserts that

> We are still, and probably even to a much larger and deeper extent, *fremde* [foreign] to nature. Either we have not achieved the ability to communicate with *her* [emphasis mine], or that has been a pathological communication. Or—as it is assumed by many contemporary natural scientists, the communication between humans and nature is nothing but a metaphor. (1998, 344)

That the communication has been pathological is an apt characterization, particularly since most in Western society do not listen to nature, but rather talk at nature and with fellow humans about how to manage *her* to our economic benefit. We are still designating an "us" and an "other," as though we are viewing a landscape of which we are not a part, instead of actually being part of the same semiosphere. However, just like the use of the term "anthropo" in the proposed designation of the epoch, the "we" to which Kull refers represents only a part of humankind: some humans do perceive and practice a two-way, respectful communication with nature, in the same way as cells in a body communicate with one another. This failure to identify with nature, and even to consider ourselves transcendent to nature, has resulted in a pathological relationship like that of a sociopath who acts as if others beyond the self do not count, and it does not matter if they come to harm. I contend that the term "Anthropocene" does not allow room for such critique.

Others have tied the exploitation of the poor with the exploitation of nature. Halliday (2001) notes that classism springs from the same source as speciesism. The hegemony of humans over nonhumans is comparable to the hegemony of some humans over all others (Halliday 2001, 198). While we have begun to confront the ideologies of sexism and racism, the ideologies of speciesism, classism, and growthism are deeply embedded in our language. The name Anthropocene promotes this worldview. Words like "sustainability" and "sustainable growth" are our attempts to put a positive spin on our current situation. Other language, repeated and used to the point of becoming a silent, automatic mantra, fixes our perspective and notions of what is good:

> Everything here, and in countless other texts repeated daily all around the world contains a simple message: growth is good. Many is better than few, more is better than less, big is better than small, grow is better than shrink, up is better than down. Gross National Products must go up, standards of living must rise, productivity must increase. (ibid., 192)

These socio-economic priorities are anthropocentric. Naming this time the Anthropocene provides a clear endorsement of these values for the privileged humans who will benefit. This myth of an always upward trajectory has the ring of destiny and inevitability, or at least of rightness, but history has proven this is not so many times over. Production itself, Halliday asserts, is a clever word trick. Nothing is produced in production; things simply change from one form to another generally with bad effects on the environment.

## COGNITIVE ETHOLOGY AND SPECIES EXCEPTIONALISM

There are scientists who believe that scientific thinking only works when the scientist sees her/himself as a part of nature. Ilya Prigogine writes that "nature cannot be described 'from the outside'" (1984, 300). Being a scientist is not a spectator sport. Communication with nature is done through description and this "description is dialogue, communication, and this communication is subject to constraints that demonstrate that we are macroscopic beings embedded in the physical world."

Cognitive ethology is a field of science that has blossomed recently and has made profound discoveries by following this model for scientific investigation. A cognitive ethologist embeds her/himself in the environment of the animal beings they wish to study. They are observers, but not disconnected spectators who do not interact with their subjects or who compose artificial environments in order to have a "controlled" experiment that could be replicated in a human scientific laboratory. In his book *The Animal Manifesto,* Bekoff writes, "How we study animals influences what we find—this is the observer effect. Too often, scientists take animals out of their natural environments and communities, place them in sterile cages or labs (where they may be held in isolation for years) and come up with all sorts of misleading conclusions about their emotional and cognitive capacities" (2010, 67). Ethologists are coming into their own as dedicated scientists who read the signs of other species and thereby gain insight into the semiotic systems of animal beings. What has resulted is a growing body of evidence attesting to the agency, intelligence, and emotional lives of animals other than ourselves.

Marc Bekoff, a professor emeritus of ecology and environmental biology at the University of Colorado and a cognitive ethologist, argues against the concept of human exceptionalism, a necessary assumption for the Anthropocene, and instead advocates *species* exceptionalism. In his article "Animal minds and the Foible of Human Exceptionalism" he says: "Speciesism doesn't work because it assumes human exceptionalism and also because it ignores within-species variation that often is more marked than between-species differences" (2015, np). This is the same type of observation made by those researching the reality of differences between the human genders: there is more variability within a gender than between the whole of one gender compared with the whole of the other. Speciesism doesn't work for the same reason sexism doesn't work. The presumed differences between the genders, legitimized by philosophy and religion, were effective at sanctioning male dominance and control over females through millennia. Bekoff continues by calling human exceptionalism "a hollow, shallow, and self-serving perspective on who we are."

My argument for the dangers of the term Anthropocene is similar in how it could direct ethical reasoning. A core consideration of ethics is "who counts?" In essence, this means who (or what, depending on definitions) merits our consideration in making decisions? What characteristics or capacities must an individual possess to be taken into account? Faulty assumptions and approaches to inquiry have given short shrift to other-than-human beings by failing to recognize animal capacities and cultures. Most current environmental philosophy agrees that animals deserve moral consideration because they can think and feel, and among those feelings are altruism and compassion. There are innumerable examples of animals exhibiting intra-species compassion and inter-species compassion. For example, BBC news reported how two whales were saved by a dolphin. A group of humans had tried to get two stranded whales on a New Zealand coast back into the water, but after repeated efforts they were exhausted and gave up. Then a dolphin arrived and appeared to successfully communicate with the whales and persuade them to return to the ocean. The conservation officer, who observed the struggle and resolution, said, "'I don't speak whale and I don't speak dolphin . . . but there was obviously something that went on because the two whales changed their attitude from being quite distressed to following the dolphin quite willingly and directly along the beach and straight out to sea'" (Bekoff 2010, 86). In the article "Minding the Animals: Ethology and the Obsolescence of Left Humanism" Steven Best (2011) writes,

> Many animals have a clear sense of morality, justice, and fair play. Great apes, elephants, wolves, whales, dolphins, hyenas, rats, and mice are capable of a wide range of moral behavior. Many believed that only humans shared food, but bonobos and chimpanzees also enjoy this ritual. Animals are not merely self-interested, unreflective, non-feeling beings locked into a violent and competitive struggle for survival with one another, they have an empathetic and altruistic side.

A post-Cartesian view of animal beings and an understanding of their capacities should debunk the notion that human beings and animal beings are poles apart: we have a lot in common, including qualities typically considered to be the better traits of humans.

Derek Jensen (2013), in the essay "Age of the Sociopath" writes, "The word 'Anthropocene' attempts to naturalize the murder of the planet by pretending the problem is "man," and not a specific type of man connected to this particular culture." I argue that one of the reasons the term "Anthropocene" is inappropriate is that it includes all of humanity while it is actually humans of industrialized, capitalistic societies who blithely silence the greater part of the semiosphere with the great noise of our technology and tunnel-visioned drive to use the Earth as if we are innocently ransacking an unoccupied planet, or at least one not occupied by anything of import other than humankind or with anything to communicate. The use of this designation would lead us to perceive the Anthropocene as a natural geologic process, one in which all humans are simply being their natural selves and thus have no guilt or agency in the state of the planet today.

## WHICH HUMANS? THE CAPITALIST/MILITARISTIC CONNECTION

Native Americans, whose numbers and cultures were decimated in the genocide perpetuated by the invaders and later the occupiers, are not the only humans who are unfairly included in the currently proposed "Age of Humankind." While numbers of the indigenous have fallen, the population of humans around the world has soared, especially in the so-called developing world. Population growth is generally held to be a primary contributor to climate change. If this were true, and areas with the greatest population growth correlated to the areas with the greatest impact on climate change, then there would be reason to hold human beings as a species accountable for climate change. However, a significant section of humankind had nothing to do with the industrial revolution, the steam engine, or the atomic bomb. As Malm and Hornborg observe, this same

> significant chunk of humanity is not party to the fossil fuel economy at all: hundreds of millions rely on charcoal, firewood or organic waste such as dung for all domestic purposes. [. . .] Their contribution is close to zero. Moreover, over 2 billion people, or one third of humanity, have no access to electricity, and so, in the words of Vaclav Smil, 'the difference between energy consumption between a subsistence pastoralist and an average Canadian may easily be more than 1,000 fold.' (2014, 65)

As Malm and Hornborg suggest, this 1,000-fold discrepancy in a human's environmental impact, depending on where and when the individual was born, makes it questionable to consider the entire species of humans culpable and the referent of the term Anthropocene.

Naomi Klein, author of *This Changes Everything*, lays the blame for the relatively sudden negative changes to the climate on capitalism. She writes, "We are stuck because the actions that would give us the best chance of averting catastrophe—and would benefit the vast majority—are extremely threatening to an elite minority that has a stranglehold over our economy, our political process, and most of our major media outlets" (Klein 2014, 18). Invoking the values and actions of capitalists as the drivers of environmental degradation limits the culpability of humans to a subset. While a human social system is still the cause of our current environmental situation, it is only one type of human social system that is at fault. The successful capitalists are wealthy, the wealthy are powerful, and theirs is a philosophy that privileges their profit above all else.

This limited scope, however, does not make the problem any easier to solve.

Most Westerners and many others around the world are not likely to contest this hegemony, because we are wedded to the fossil economy and the idea of humans as masters of the world. And it is not easy or cheap being masters of the world. The second Iraq war, for example, produced at least 141 million metric tons of carbon dioxide (Bast 2008).

## HUMAN EXCEPTIONALISM

To seriously consider the current epoch the Anthropocene, one would necessarily need to view humans in isolation as the most important feature of current time, historical time, and future time, on a geological timescale. One would need to hold the assumption that humans are exceptional. Human exceptionalism is, in fact, a trending, popular notion. Not surprisingly, humans themselves are the genesis of the designation and doubtless the sole species to embrace the concept. Such a "Narcissus" reference point is an apt way to set humans as the gods and monsters of the Earth, and thus it serves the purpose of those who deplore the state of the environment at the hands of humans as well as those who see humans as omnipotent and unsurpassed in their capacity to control, dominate, and manage. Max Oeschlaeger points out, however, that human presumptiveness is not conducive to solving our current problems on the Earth. He writes, "what the new conservation proposes is essentially an audacious, global scheme of managing planet Earth, as the Earth is nothing more than so many parks, gardens, impounded rivers, forest and agricultural fields" (Oeschlaeger 2014, 245). He goes on to assert that the failure of Biosphere 1 and 2 demonstrates our inability to create viable systems, and that indeed there is no example of a human-controlled ecosystem, such as with forests and fisheries, that has been sustainable. Hence the idea that we as gods can create our own Garden of Eden is not a viable goal.

Although the term "human exceptionalism" has only recently become popularized, the concept itself has a long history. The Discovery Institute (2015) holds this idea as central. On its homepage are the words: "Privileged Species: How the Cosmos is Designed for Human Life." A deconstruction of this catchy phrase could be along the lines of "How Humans Decided the Cosmos was Theirs at the Peril of Everyone Else." At the foundation of this worldview is the Judeo-Christian tradition. It is here in the Genesis myth of creation that adherents claim the religious and philosophical basis for the primacy of humans is established. Depending on the version, humans are assigned a role of dominance or stewardship over the Earth and all other sentient beings. This is also the myth that gives humans the right to subjectively name all other life, and by so naming, to define, judge, and appropriate. Lynn White traces the effects of the Judeo-Christian theology on the environment in his seminal essay, "The Historical Roots of our Ecological Crisis" (1974).

Contrary to the predominating philosophical stance of the Church through the ages, the current pope, Pope Francis, has been promoting a fierce message favoring the environment and the poor. His encyclical (2015) has been causing great irritation within the oil industry because he is preaching taking the long view with nature, protecting and caring for the Earth, and a socialist message of greater economic equality. He has also quoted a fourth-century bishop when he called corporate greed "the dung of the devil." "Devil's dung" is a punchy moniker, assigning an immortal a mortal biological function. This might metaphorically be an appropriate name for this period.

Admittedly, the name Anthropocene does point to our isolation from the natural world and I assert the suffering we therefore cause ourselves. Psychologists have a name for this: Nature deficit disorder. But this term is inadequate because it generally refers only to human children and not to adults, as indeed it should. As David Orr writes, "[T]he average person has come to recognize over 1000 corporate logos but can now recognize fewer than 10 plants or animals native to his or her locality" (1999, 696). Perhaps one reason for focusing on children is the hope that they are still malleable and might be the ones who will grow up to translate for us. As an experiential component of a course I taught called "Valuing Wilderness," I took a group of college students on an expedition to the Boundary Waters, a wilderness area bordering Minnesota and Canada. At first I thought I might succeed in embedding them in the wilderness because their mobiles did not receive a signal at our campsite on an island. This, however, did not change their human-centric values. They chatted about their jobs and going shopping at the mall. Only one student had the temerity to leave the island for an hour. One afternoon I left the island to canoe in the area, and when I returned, I saw them all standing on the shore, looking out, like the shipwrecked characters of the American sitcom *Gilligan's Island*. I had considered taking the students singly, or even in pairs, to camp for a night on a different unoccupied island, but I realized that they would have talked about everything except the nature in which they were embedded. It would have been like an empty stage set for them, where they would have fretfully waited for deliverance. It might have resembled *Waiting for Godot*, except that the stage would have been empty only in their minds.

Fundamentally, we have a deep, psychological, spiritual, and ethical problem: we do not know what we do not know because we do not read the signs. To ransom ourselves from the Anthropocene, we need to pull away from our reflection, sit by the pond with its many beings as our companions, and inhale the air in which we sit and on which we depend. We need to listen, to see, to smell, and to touch with a wider sense of self, with compassion, empathy, and respect for all who live here within a living Earth.

As Kathleen Dean Moore of Oregon State University writes, "[W]e should use words cautiously. Words are powerful, magical, impossible to control. With a single misguided phrase, they can move a concept from one world into another, altering forever the landscape of our thinking. It's essential that we get this straight now" (Moore 2013, np).

# TWO

# Dangerous Intersubjectivities from Dionysos to Kanzi

*Louise Westling*

In the midst of what scientists and journalists are calling the Sixth Extinction, we face an urgent need for a reevaluation of human relationships with other animals, with which we share an uncertain fate on a radically diminished planet. Can we learn to live together in this future, or will humans destroy most other animals and thus probably ourselves? This chapter considers cultural understandings of human/animal relations from the earliest art to contemporary primatology, in order to gesture toward a new ethics. Paleolithic and Bronze Age painting and sculpture depict a world view dominated by powerful animals. Both the *Epic of Gilgamesh* and Euripides's *Bacchae* imagine catastrophes caused by arrogant refusal to appreciate the intertwining of humans with the wild forces and creatures in the natural world. Species arrogance and indifference to all other life continue to lead *Homo sapiens* to overwhelm and destroy wild communities and landscapes all over the planet. This discussion will examine both the common semiotic scaffolding that links our biology and behavior with those of other animals, and also the differences that must be respected as habitats shrink and climate change caused by human activity forces migrations, extinctions, and radically different interspecies relationships. Particular focus will be given to efforts to create "bi-cultural" spaces and "enlightened" zoos where humans and other primates such as chimpanzees and bonobos can learn to "talk" and cooperate. The work of primatologists Frans de Waal, Sue Savage-Rumbaugh, and Frances White will be placed in dialogue with Franz Kafka's "A Report to an Academy" and Karen Joy Fowler's novel

*We Are All Completely beside Ourselves* to think about excessive human intrusions into the lives of other animals and call for a new biosemiotically informed ethics.

Kafka's urbane chimpanzee Red Peter addresses an unnamed academy about his transformation from an ape five years earlier into a "person" who successfully adopted "the way of humanity."

> . . . your life as apes, gentlemen, insofar as something of that kind lies behind you, cannot be farther removed from you than mine is from me. Yet everyone on earth feels a tickling at the heels; the small chimpanzee and the great Achilles alike. (Kafka 1959, 174)

The dilemma which Kafka imagined in "A Report to an Academy" has become a real and disturbing one in recent decades, though humans are primarily responsible rather than apes who seek to cross the species boundary as "a way out" of their cages. Well-meaning and illuminating as Ape Language Research and other interspecific efforts at communication may be, sad consequences have fallen upon such famous primates as Nim Chimpsky, Washoe, Koko, and Kanzi; Alex the Parrot; and cetaceans studied by human scientists eager to try to create linguistic or semiotic relationships with them. Jane Goodall said recently that chimpanzees experience psychological trauma when kept with humans as infants and young family members but then have to be removed when they grow too strong and dangerous for their human companions. "They're stuck between two worlds. They've never learned to be a chimpanzee and they can never become a human" (Goodall 2015).

Wittgenstein turns out to have been wrong to some extent in his famous claim that if a lion could speak, we could not understand him. Anyone who lives with a dog or cat or who works with animals such as horses and sheep knows that trans-species communication happens all the time, by gesture, bodily movement, and audial or verbal signals. We could not live with domesticated animals otherwise. Such relationships have evolved with all our various animal kindred over millions of years, but communication is only partial and is primarily limited to domesticated animals. People have longed to be able to speak with other animals more fully, including wild ones, expressing in myth, oratory, fantasy, poetry, and fiction a sense that more complete communication might once have been possible. However, during the several hundred years since Descartes, most Western philosophers and scientists have insisted that only humans have language, conscious thought, and culture. Montaigne, Hume, and Darwin were influential dissenters, but they were overruled by the powerful inertia of humanist exceptionalism. We are all aware that we have reached a crisis, which Jacques Derrida defines in *The Animal That Therefore I Am* when he speaks of the horrific developments of "zoological, ethological, biological, and genetic forms of *knowledge*, which remain inseparable from *techniques* of intervention" such as industrialized meat "production" and genetic experimentation (Derrida 2008, 25). However, as Dominique Lestel indicates in "The Withering of Shared Life Through the Loss of Biodiversity," attitudes among scientists and the public have been gradually changing (Lestel 2013). Some of the ethological and linguistic work mentioned above, as well as major changes in primatology (see de Waal 2001; Haraway 1989; Fossey 1983; Smuts 1987, 1999) and the rapidly developing theoretical field of critical animal studies have been major causes of this change. Twentieth-century philosophers such as Wolfgang Köhler, Maurice Merleau-Ponty, and Derrida himself anticipated or contributed to present concerns, and leading anthropologists such as Tim Ingold and Philippe Descola have been calling for an end to dualistic thinking that separates humans from other animals. But species arrogance and long habits of power encoded in scientific practice have blinded many of the most progressive researchers to the necessary limits of our intrusions into the lives of other animals. My purpose is to set the present crisis of human/animal relations within the long history of our species, and then to look closely at recent ape language research (ALR), in order to gesture toward a new biosemiotic ethics of relationality and restraint. This approach was implied in Merleau-Ponty's Nature lectures of more than fifty years ago and has been developed through increasing evidence of genetic similarities and evolutionary relationships among animals in biological sciences of recent decades, as Jesper Hoffmeyer demonstrates in *Biosemiotics: The Signs of Life and the Life of Signs* (2008). Deep cultural contexts will be a good place to start, with a brief look at a few ancient indications of the evolving human cultural representations of animals. Then I will briefly review both the common biological inheritance and semiotic scaffolding that links our bodies and behaviors with those of other animals, but also the crucial differences that must be respected as habitats shrink and climate change forces migrations, extinctions, and the likelihood of radically different interspecies relationships in the future. Primate studies offer a particularly revealing example, specifically Ape Language Research and efforts to create "bi-cultural" spaces and "enlightened" zoos where primates such as chimpanzees and bonobos can learn to "talk" and cooperate with researchers. Whether or not we should call our historical era the Anthropocene, there can be little doubt that humans have caused most of the present crisis by their reproductive success, by their aggressive technologies, and also by arrogant curiosity. Thus it is imperative that we radically reevaluate ourselves and change our behavior toward other animals, as well as toward the wider environment. That means seeing ourselves as fellow animals intertwined in a shared world, yet at the same time constrained by species limitations.

## ARCHAIC REPRESENTATIONS OF HUMAN/ANIMAL RELATIONSHIPS

The earliest art produced by modern humans expresses a profound awareness of a vast community of exuberant, powerful animal life. We are all familiar with the magnificent cave paintings from Lascaux, Altamira, Chauvet, and other sites in France and Spain. Whatever their exact purposes for those ancestors of 30,000 to 15,000 years ago, we cannot doubt their intense attention to the panoply of fellow creatures from bison and aurochs to lions, birds, fish, reptiles, and plants. Human images are rare and are dwarfed by those of the huge animals around them. Alexander Marshack's microscopic and infra-red photography of mobilary art (chiefly bone carvings) and cave paintings revealed a complexity of sophisticated detail and attention to animal behaviors and seasonal plant life, as well as lunar calendrical notations, which had not been seen before his work in the 1960s (Marshack 1972, 1975). Among Marshack's microscopic images of bone carvings are several he calls "sorcerers"—apparently shamanistic or ritual hybrids of male human figures with heads of stags or bison, or strange masks (Marshack 1972, 273-274), that suggest concepts of human-animal synthesis or transformation. Such ritualistic images are familiar from many later tribal hunting cultures such as Native American tribes of the Pacific Northwest and Hopi, Navajo, Pueblo, and Lakota Sioux among others. They persist in traditional celebrations where dancers wear the heads of stags, bison, turkeys, or even butterflies (https://www.youtube.com/watch?v=iW10cp223So; https://www.youtube.com/watch?v=AqdSP-bIBSU; https://www.youtube.com/watch?v=D9jreJdXQP8).

What is less known in Western academic circles is the existence of similar representations in Asia, particularly in rock art of the Mongolian and Russian Altai. Esther Jacobson-Tepfer's recent book, *The Hunter, the Stag, and the Mother of Animals* (2015), details two issues that relate directly to the concepts I have introduced above. On the one hand, the palimpsests of animals overlaid but without any psychological relationship to each other indicate that the complex overlays we see, for example, in the French Chauvet Cave, may have a fundamental universal significance, perhaps referring to the interpenetration of all life. On the other hand, within a long chronological span from the pre-Bronze Age through the early Iron Age, Jacobson-Tepfer identifies half human-half animal figures—what she calls liminal beings—that indicate the existence of a being both generative and destructive. These figures appear on great standing stones in the Minusinsk steppe and on rock-pecked panels within the Altai. Horned, bird-like, or moth-like, these Animal Mother figures gradually devolve in the Bronze Age under the impact of emerging narrative structures. The original Animal Mother becomes a birthing woman—the source of animal life—and finally a realistic female figure representing the Mistress of the Mountain, the one who assures the hunter of success (Jacobson-Tepfer 2015, 4-25, 59-64, 77, 88, 93, 97). This material seems to point to the same fundamental characteristic of Paleolithic European cave art: an abiding belief in the interconnection of human and animal life.

We cannot know very much about the specific cultural meanings of these early art forms, though many efforts have been made and debated (see, for example, Leroi-Gourhan 1967, 106-150; Marshack 1972; Jacobson-Tepfer 2015, 368-369). Jacobson-Tepfer discusses the challenges of interpreting the meanings of ancient artifacts in light of oral traditions and later cultural behaviors and productions, because of shifting emphases "under a great range of social, environmental, and cultural pressures" (2015, 316). Nevertheless, cultures evolve by adapting and translating traditional materials for the needs of particular historical moments, and it is tantalizing to consider their relations to modern understandings of animals. One thing is clear—representations and practices focused on the power of animals dominated most of human cultural experience from at least 35,000 years ago until quite recently. Many are still performed in traditional folk festivals around the world. I have suggested elsewhere that these art forms and ritual practices embody ecological understandings and experiences that lie just beneath contemporary world views based on Enlightenment philosophy and modern science and technology. The older forms still make up the deepest levels of our understanding as sedimentations of an ancient ecological imaginary that nourishes all cultures (Descola 2013, 70-71; Westling 2016). By "imaginary" I mean the French sense of the word as a noun, which philosopher Annabelle Dufourcq defines in terms of Merleau-Ponty's enlarged concept of reality that is not based on traditional oppositions such as real/ideal, real/imaginary, real/possible. Instead, the imaginary, the realm of dream and art, is a vivid, fluctuating, intense, shimmering, ambiguous part of the real that is full of possibilities but also haunted and nightmarish (Dufourcq 2014, 709). As anthropologist Deborah Bird Rose learned from her aboriginal teachers in North Australia, "the living world is more complicated, less predictable, more filled with transformations, uncertainty, and fantastic eruptions of life's mysteries than is allowed of in ordinary [modern] thought" (Rose 2011, 5–6). Central to these understandings are feelings of profound connection with animals that began to emerge again in Darwin's work and in ethology in the nineteenth century and are now returning in contemporary literature, film, and digital media, as well as theoretical reconsiderations of human/animal relationships as we find ourselves in an increasingly destabilized global environment. Annabelle Dufourcq asserts that Merleau-Ponty's philosophy brings Husserl's notion of the crisis of Modernity into focus "as a dreadful manifestation of the irrational heart of existence" in which human control seems lost and we are returned to an uncertain, open, and enigmatic reality "as if man had to undergo again the Dionysian ordeals of Greek mystery plays" (2014, 709).

## BIOLOGICAL RELATIONSHIPS AND SEMIOTIC SCAFFOLDING

Radical challenges to human exceptionalism have come from embryology and recent genomic studies that reveal bodily relationships with all animals in much more intricate, molecular terms than Darwin knew but which support his assertions in *The Descent of Man* (1981, 34–53). We have ancient genes, many of which we share with other animals and with plants. Some plants have much larger and more complex genomes than humans do. Ninety percent of each human body is composed of bacteria and viruses, some of them being the same symbiotic microbial organisms as those inhabiting other animals (Hoffmeyer 2012, 180–182; Wenner 2007; Margulis 1998, 65; Lingis 2003, 165–167; Rose and Oakley 2007). Each human is thus a microcosmic ecosystem working as large collective being, containing sedimented genetic memory, and symbiotically interacting with other organisms all the time. Indeed, as Jesper Hoffmeyer explains, "every single life-form in existence today has, lodged inside its genetic material, the sinuous trail of its evolutionary past harking all the way back to the dawn of life—while it is itself busy incorporating the experiences of today into the future" (Hoffmeyer 1996, 13). Dynamically functioning within (the *innenwelt*) as well as outside each animal body (the *umwelt*) are continual meaningful communications with and responses to the environment around it. Like the biological sedimentations of evolutionary experience, ancient cultural expressions lie deep within our languages and art forms. Merleau-Ponty posited "a meaning of Being absolutely different from the 'represented,' that is, the vertical Being which none of the 'representations' exhaust and which all 'reach,' the wild Being. [. . .] and even the cultural rests on the polymorphism of the wild Being" (1968, 253). The environment of brute existence and essence is not something mysterious: "we never quit it, we have no other environment" (1968, 116-117). According to his chiasmic ontology of intersubjectivity, reciprocity, and intertwining of beings and things, vast and intricate synergies operate in intercorporeal ways within us and in our external relations with the world. "Why would not the synergy exist among different organisms, if it is possible within each?" (1968, 142) Contemporary biology and biosemiotics clearly support such a view, and as Dominique Lestel asserts, this is the very texture of life and meaning, or as Merleau-Ponty phrased it, the Flesh of the World (see also Morton 2010).

## ANXIETY IN ANCIENT TEXTS ABOUT HUMAN VIOLENCE AGAINST NATURE AND ANIMALS

In spite of these interrelationships, Lestel has suggested that Western culture is motivated by a hatred of Nature (Lestel 2013, 314). Certainly we see such hostility all around us, but as suggested above, I believe that the older attitudes continue to exist. I have discussed these matters elsewhere (Westling 1996, 2014), but here I would like briefly to return to two ancient literary works that suggest a transitional stage from the world views of our Paleolithic ancestors, a stage of deep anxiety about the desire to separate ourselves from the rest of nature. The conflation of animal and human forms in the main characters of the most ancient known literary work, the Mesopotamian *Epic of Gilgamesh*, and Euripides' tragedy, *The Bacchae*, carry into written form much older preoccupations and uneasily synthesize or combine them with the heroic energies of complex agricultural societies. Gilgamesh is a dangerously powerful king of Uruk who is likened to a rampant wild bull tormenting his subjects rather than protecting them. The gods subdue him by creating a *doppelganger* or complementary companion, Enkidu, who begins his life as a hairy creature living with the beasts of the countryside and ignorant of human kind (Dalley 1989, 53–61). When Enkidu is seduced by a harlot (or priestess of the goddess Inanna/Ishtar) and led into civilization, he and Gilgamesh meet, clash, and then become allies who share each other's qualities. However, their attack on a great forest sacred to the gods and their slaying of the wild bull of Ishtar bring retribution and death, suggesting that such heroic arrogance against wild nature and its avatars and guardians is disastrous. The Greek god Dionysos, or Bacchus, is a figure representing the powers of wild animals and plants, often appearing as a magic bull and associated with snakes and twining ivy, as well as the fertile earth (Bagg 1978, 2-3, 23, 38, 40; Dodds 1989, xi-xii, xx). Euripides's play dramatizes his revenge upon arrogant humans who refuse to acknowledge his power. In particular, he is opposed by his cousin, the young King Pentheus, who is also symbolically related to wild natural creatures, particularly snakes, but unconscious of these connections and hostile to the worship of the god. He tries to imprison Dionysos, who appears as a stranger devoted to the god, but the "man" put in chains in the stable transforms into a bull that cannot be captured. Eventually, spying upon the mysterious worship of the god's female followers, Pentheus is attacked and dismembered by them. His own mother thinks she has killed a lion and bears his bloody head back into town, not recognizing what she and the other worshippers have done. Thus transformations of human and wild animal form again express deep anxieties about denial of profound kinship with other animals. These works insist upon the crucial necessity to respect other animals and their independent life, and the tragic consequences of efforts to deny the integral human place within the biosphere.

In Yuri Lotman's definition of the semiosphere, texts exist in dynamic relationships with each other, with the culture in which they arise, and in contacts with other cultures where they are translated and absorbed in new ways. A text like *The Epic of Gilgamesh* or *The Bacchae* would be for Lotman "not only the generator of new meanings, but also a condenser of cultural memory" for its audience, thus acquiring a semiotic life. Works from other cultures and time periods (e.g., archaeological remains) are not just irrelevant museum pieces but

> are important factors in the stimulus of cultural dynamics. For a text, like a grain of wheat which contains within itself the programme of its future development, is not something given once and for all and never changing. The inner and as yet unfinalized determinacy of its structure provides a reservoir of dynamism when influenced by contacts with new contexts. (Lotman 1990, 18)

Jesper Hoffmeyer compares such structures to the biological semiotic scaffolding provided by the genomes of organisms, suggesting that literary texts and films provide "scaffolding devices for cultural development" (2008, 138).

## ECOLOGICAL CODES AND OVERLAPPING *UMWELTEN*

In a bold suggestion of a similar linkage between human cultural semiotic scaffolding and wider biosemiotic processes, Timo Maran has posited the existence of ecological codes that might be similar to archetypal structures like those defined by Carl Jung for the human psyche. These could be involved in semiotic processes taking place between different species and in biosemiotic processes at the ecosystem level. They would not be as precise as the codes within one species' communication, for example, human language or particular bird calls. Rather they would be "ambiguous and fuzzy linkages based on analogies and correspondences" as in eye-marks on moth or butterfly wings and insectivorous birds. An ecological code would have three basic properties:

- It would involve different species with different perceptual abilities, and therefore "no single individual or species would have full perception of an ecological code."
- An ecological code would develop within a particular ecological community and use "habitual semiosis, behavior and action of animals"; and such codes would be communal and disperse from their original locations.
- An ecological code would use different memory types, both cognitive and non-cognitive.

Maran explains that we can replace Jung's notion of the collective human unconscious with "interspecific" consciousness and "species-specific" consciousness. We would take "unconscious" in a Sebeokian sense to refer to nonlinguistic layers in the semiotic self. "Jung's theory is originally aimed to describe the psychological content of human species, whereas in the study of ecological codes archetypes should be widened to include umwelten of other animals as well as interspecific semiotic and ecological relations" (Maran 2012, 148-152). Kalevi Kull and Peeter Torop's discussion of "Biotranslation between Umwelten" suggests that contrary to Jakob von Uexküll's concept of umwelten as closed units (Uexküll 1982, 30-31), more recent research and theory have expanded the concept to include overlapping of umwelten, translations between them, and many other kinds of interrelations among the umwelten of different species based on recent decades of work in interspecific communication. Kull and Torop offer the example of two different species of birds feeding in a place where one can see an approaching cat and the other cannot. The first gives an alarm cry and flies away, which the other interprets as a call to also fly away. Such processes can be seen at biological levels from the individual down to genetic inheritance, which is interpreted or translated in a particular way by each individual (Kull and Torop 2011, 414-421; Kull 1998, 349-351; Tønnessen 2015, 9-12, 15-19; cf. Brentari 2015, 154). At a common sense level, one can readily see that such processes between species are everywhere apparent. British shepherd James Rebanks tells the story of a fox that tricked hunters and their dogs in his Cumbrian landscape by running in and out of holes in hedgerows and cracks in walls and gates to hinder the following hounds, then by running into a flock of sheep to confuse his scent, and finally by turning from the open pasture into a reedy bog full of strong smells. By the time the dogs had recovered the fox's scent, he was far away, beyond capture (Rebanks 2015, 213-215). This example is an even more dramatic illustration than the ordinary kind of cross-species interpretation and communication which I have elsewhere described for the activity of herding sheep with dogs (Westling 2014), in which animals understand each other's *umwelten* and devise strategies to manipulate semiotic relationships. Human experience for hundreds of thousands if not millions of years has been formed within these networks or bodily intertwinings of sedimented genetic evolutionary memory and cultural experiences with other animals. A few remaining tribal societies still understand their worlds in light of these connections where their traditional lifeways have continued (Descola 2013, 3-31; Rose 2011). Cynthia Willett reminds us that "our ancestors" survival depended on exquisite sensitivity to the subtle movements and nuanced communication of predators, prey, competitors, and all the animals whose keener sense of vision, smell, or hearing enhanced human apprehension of the world (2014, 98). Their lives also depended on cooperative relationships with fellow predators such as wolves, as well as eventual alliances with prey animals such as horses, goats, sheep, and camels who eventually became domesticated associates, or in the case of wolves, essential partners (Westling 2014; Csányi 2005; Pryor 2003). Nevertheless, Merleau-Ponty's concepts of *écart* and dehiscence emphasize the distinctions among all creatures, unbridgeable differences that exist between species, even as all are interrelated biologically and participate synergistically within the flesh of the world (1968, 147-148).

In spite of distinctive *umwelten* and semiotic habits, animals with which we have lived closely for thousands of years have learned to understand our behaviors, as we have come to know theirs. Dogs in particular can read human facial expressions better than our closest primate relatives (Hare and Tomasello 2006, 497), and Vinciane Despret asserts that the famous horse Clever Hans had an intricate and subtle understanding of human gestures and facial expressions. This allowed him to shape the collaboration that resulted in his correct answers to mathematical and spelling questions, and his discrimination among colors and musical tones. "The practice was not on the questioner's side only; Hans was teaching them what made him move. Hans the horse was as much leading them as the humans were leading him" (Despret 2004, 116). Vicki Hearne tells the story of a horse who understood the slightest voice and body signals from her injured rider to win an intricate Grand Prix jump-off trial (Hearne 2000, 118-121). But these are domesticated animals that assent to cooperaton with humans in spite of very clear species differences. Animals living independently, outside human-controlled communities and environments—those we call "wild"—have not made those tacit agreements over millennia of cooperation, and they remain unpredictable, dangerous, unwilling to participate in training practices that allow humans to work effectively with such animals as dogs, horses, sheep, elephants, yaks, water buffalo, and camels (Hearne 2000, 18-41). "Wild" animals must be caged and coerced, often with punishment, to be worked as circus performers or to be controlled for scientific experimentation. But the human dream of contact and understanding with wild animals continues to draw visitors to zoos and animal documentary films, as well as to motivate attempts to cross species barriers and find methods for direct communication.

## APE LANGUAGE RESEARCH AND EFFORTS TOWARD "BI-CULTURAL RELATIONSHIPS/SPACES"

In the case of our closest primate relatives—the gorilla, chimpanzee, bonobo, and orangutan—humans have always been fascinated by their close physical similarities, though also perturbed by their differences from us and their essential wildness. In the past one hundred years, this fascination has returned in force. As Donna Haraway and Frans de Waal explain, European and American scientific work with chimpanzees and other primates has tended to follow the Cartesian scientific practice of rigidly disciplined experimental regimes in controlled laboratory settings, in which apes and monkeys are treated as objects (Haraway 1989, 115-132; de Waal 2001, 117-119, 188-212; Lestel 2011, 378-379). Behaviorist scientists carried this kind of practice to extremes with horrible consequences of injury and death (de Waal 2001, 88). Much experimentation has been focused on medical research that is still practiced but increasingly protested, in which animals live in cages and are subjected to surgical and pharmaceutical interventions (see Haraway 1989, 121-123). Some experimental studies were exceptions, such as those of Wolfgang Köhler from 1913-1919 on chimpanzees in the Canary Islands, during which he investigated the cognitive abilities of his subjects. Frans de Waal describes how traditional familiarity with primates and lack of philosophical traditions separating humans from other animals fostered different approaches among Asian researchers, such as Japanese primatologist Kinji Imanishi, who soon after World War II began studying wild monkeys and apes in their natural habitats and discovered complex cognitive and cultural abilities not previously understood (de Waal 2001, 110-126). British archaeologists Lewis and Mary Leakey, convinced by their archaeological discoveries of early hominins in Africa that there must be close relationships between these early ancestors and the chimpanzees, encouraged Jane Goodall to study chimpanzees in their natural African habitat in the 1960s. We all know about the revolution in attitudes toward chimpanzees, bonobos, gorillas, and orangutans that followed, with increasingly rich discoveries about ape cognition and behavior by primatologists in the past fifty years (de Waal 2001, 117, 188-189; Haraway 1989, 129-146).

During this period Ape Language Research (ALR) began, with the famous studies of the chimpanzees Washoe and Nim Chimpsky, Koko the gorilla, and Kanzi the bonobo. Haraway explored the motivations behind these efforts, concluding, "Surrogates, rehabilitants, language students, and adopted children: apes modeled a solution to a deep cultural anxiety sharpened by the real possibility in the late twentieth century of western people's destruction of the earth" (1989, 132). As all of these projects are quite famous, I will not rehash them. Instead I want to concentrate on the problems associated with these intelligent animals' lives, and negative consequences that are not widely known. First, however, we must admit that valuable knowledge came from the work with these apes, who demonstrated to varying degrees that they could learn to communicate in something like human language up to the level of a two- or three-year-old child. There have been criticisms of the claims of Allen and Beatrix Gardner and Roger and Deborah Fouts who taught Washoe American Sign Language, Herbert Terrace who worked with Nim Chimpsky, Francine Patterson who communicates in sign language with Koko, and Sue Savage-Rumbaugh's use of lexigram symbols and spoken English to "talk" with the gifted bonobo Kanzi (see, for example, Cheney and Seyfarth 2005; Penn, Holyoak, and Povinelli 2008). Whether or not these primates have used fully human language, there is no doubt about their successful communication with human researchers. But serious problems have come to light concerning the conditions under which these discoveries were made, the sentimentalization of some of the projects, and the sad outcome for most of the apes involved. A recent article in the online *Slate Magazine* offers a behind-the-scenes look at the present moribund state of Ape Language Research that has declined into "a surprisingly dramatic world of lawsuits, mass resignations, and dysfunctional relationships between humans and apes" (Hu 2014).

Vicki Hearne movingly describes witnessing the close confinement of the chimpanzee Washoe in her later years. Because of her strength and unpredictability, she could not spend time outside her cage except when accompanied for a walk by two male trainers with leashes, a tiger hook, and a cattle prod. She was very dangerous physically, having broken one handler's kneecap and bitten others in the past (Hearne 2000, 18-41). Up until the age of five, she had lived in circumstances as close to those of a human child as possible, with clothes, toys, furniture, and constant human companionship, but after that she was moved into more constrained quarters at a primate center at the University of Oklahoma. She died at the age of 42 in a primate experimental facility at Central Washington University. A far worse case was that of the chimpanzee Lucy, raised from infancy as a human child by Maurice Temerlin and his wife. She became increasingly destructive and difficult as she matured and was taken to a chimpanzee rehabilitation center in Gambia. In spite of having a graduate student from the University of Oklahoma working to help her assimilate with the other chimpanzees, she never accommodated to them, was deeply depressed, and died a grisly death at unknown hands ("Parent Trap" 2010).

The famous Nim Chimpsky's unfortunate life was documented in a 2011 film by James Marsh (Marsh 2011; see also Kappala-Ramsamy 2011 for an example of newspaper coverage).

Happier have been the lives of the gorilla Koko and the bonobo Kanzi. Koko lives in a special facility that is essentially a large cage in a reinforced trailer, though she seems to have a rich environment and not to need the kind of constraints or coercive protections for human associates that are required with the more dangerous chimpanzees. Her life seems to be much more comfortable, enjoyable, and stimulating with constant human companionship and cognitive enrichment. But Koko has not lived as a normal gorilla since the age of one year and has rejected opportunities for mating with male gorillas that have been brought to reside in adjacent cages. Former employees of the Gorilla Foundation have charged that she is overweight because of eating an inappropriate diet of human foods and that she almost never exercises but instead spends most of her time sleeping or watching videos in her trailer. Scientific critics have charged that most of her linguistic activity is prompted by questions from her trainer Francine Patterson, who translates for her, laughs at her mistakes, and keeps repeating questions until she receives the response she wants. Few rigorously scientific studies have been produced by Patterson and her associates (Hu 2014).

Kanzi's training has produced the most complex and successful effort at Ape Language Research, but its successes are long past. Primatologist Sue Savage-Rumbaugh attempted to set up a bi-cultural space in which Kanzi and his sister Panbanisha could interact with human researchers who shared some of their bonobo culture and activities. Savage-Rumbaugh introduced Kanzi to lexigram boards with which he learned more than 400 symbols for English words. He also developed a remarkable ability to respond to spoken English. Savage-Rumbaugh and her associates would help the bonobos make "nests" out of bedding in the evening, they would participate in grooming, and they learned to use some bonobo sounds and gestures. Kanzi adopted many human gestures and habits, learning to point to indicate directions and to read simple maps of the woods around his cage area (Savage-Rumbaugh 1998, 3-74). In a TED talk, Savage-Rumbaugh showed a video in which as an adolescent bonobo, he went for a walk with her in the woods, made a campfire, roasted and shared marshmallows, and then doused the fire with water when asked to do so (2004). Unlike Koko, Kanzi has always lived with other bonobos, including his adoptive mother Matata and a sister Panbanisha. He has fathered a number of offspring. But like Koko, he lives a caged life, carefully controlled by humans, though he does not seem as physically dangerous as an adult chimpanzee. He eats too many human foods and is seriously overweight. Frans de Waal said recently that he was receiving as many as five Frappuccinos a day before Savage-Rumbaugh was relieved of her duties as director of the private Great Ape Trust (now the Iowa Primate Learning Sanctuary) where he and some four other bonobos live in Iowa (de Waal, personal conversation, 10 March 2015). Employees of the institution had written complaints to its governing board a number of times about unsafe and unhealthy conditions, and raised serious concerns about Savage-Rumbaugh's directorship. Scientific respectability has not been restored in Kanzi's treatment. His life, like Koko's, seems to have become a kind of entertainment or circus act, as can be seen in numerous You Tube videos and the availability of public tours of the Iowa Primate Learning Sanctuary. As recently as the summer of 2014, Kanzi judged the dessert contest at the Iowa State Fair, presumably tasting many fruit pies and cakes to decide the winning entry (Wong 2012; Hu 2014).

One distinguished senior researcher in ape/human cognition told me that Savage-Rumbaugh became confused about her own identity, coming to think of herself too much like a bonobo. This possibility is eerily anticipated by Kafka's Red Peter when he describes what happened to his first teacher: "My ape nature fled out of me, head over heels and away, so that my first teacher was almost himself turned into an ape by it, had soon to give up teaching and was taken away to a mental hospital" (Kafka 1959; 183). Such may be an extreme description of blurred identities between apes and their human associates, but both Savage-Rumbaugh and Kanzi have endured painful changes in recent years. Kanzi's life has been completely interrelated with humans in a specialized environment, so that he must have experienced great sadness when he lost his central human companion at the age of thirty-four when Savage-Rumbaugh departed from the Great Ape Trust.

We cannot enter the world of even a linguistically gifted bi-cultural chimpanzee or bonobo or gorilla to know exactly what such psychic damage might be like, but novelist Karen Joy Fowler has attempted to suggest its intensity and sadness in her recent novel, *We Are All Completely Beside Ourselves* (2013; see also McAdam 2014). Basing her plot on circumstances from a combination of early Ape Language Research efforts, chiefly the Kellogg experiment with Gua in the 1930s and the Temerlins' work with Lucy in the 1960s and 1970s, Fowler tells the story of a young woman who spent the first five years of her life with a chimpanzee sister. Epigraphs from Kafka's tale of Red Peter precede the main sections of the novel, so that a talking ape seems to introduce the events to follow. But the consciousness of the story is limited to its human protagonist, Rosemary Cooke, the daughter of a psychologist at Indiana University. In their house his team of graduate students collected volumes of data comparing the physical, cognitive, and linguistic abilities of the young female chimpanzee Fern and the young human girl. Because Fowler withholds the identity of the sister who disappeared from the family at the age of five, the first quarter of the novel is Rosemary's mysterious description of her own psychological distress as a university student who does not get along well with other people and who is haunted by her damaged family and by grief for her lost sister. Thus readers are tricked into anthropocentric assumptions about Rosemary and Fern's close relationship. When we learn that Fern is a chimpanzee, those human projections continue to follow the unfolding tale of the two young females' playing together, creating mischief around the house, and the eventual destructive behaviors that cause Fern to be taken away to a primate center where she spends the rest of her life in a cage. Fowler places the ape facility in a South Dakota university town similar to Kanzi's primate "sanctuary." At the end of the novel, when Rosemary and her fugitive animal activist brother have forced their parents to admit what happened to Fern, Rosemary and her mother go to live near Fern's institution so that they can try to restore the bonds that had been so disastrously broken. In this revelatory conclusion, Rosemary confides to readers her own "Monkey Girl" identity and the impulses she has struggled to hide since childhood. The novel ends with descriptions of her mother's and her slow and painful reunion with the adult Fern who gradually recognizes them and begins again to communicate with sign language, but who remains forever behind a thick glass partition, impossible to touch. Rosemary restores contact with her chimpanzee sister by pressing a red poker chip against the window and signing her name to Fern. Colored poker chips had been the tokens for their games as little children, as Fern remembers at once. She signs her name back to Rosemary, who can't tell "if she'd remembered me or was simply being polite." Rosemary confesses both her incomprehension and her profound identification, trapped in a middle space between species and robbed of what had seemed a full sibling relationship in their very young childhood:

> I didn't know what she was thinking or feeling. Her body had become unfamiliar to me. And yet, at the very same time, I recognized everything about her. My sister, Fern. In the whole wide world, my only red poker chip. As if I were looking in a mirror. (2013, 308)

## CONSEQUENCES AND BIOSEMIOTIC ETHICS

Kafka's story about the chimpanzee Red Peter and Fowler's dramatization of a family's attempt to raise their daughter with a chimpanzee sister are only human fictions, yet they poignantly help us to imagine the damage caused by experimental attempts to know these close yet distant cousins in the strange kinship described by Merleau-Ponty in his Nature lecture on "Man and Evolution: The Human Body." That kinship is both biologically real and at the same time illusory in the sense that evolution in different (though similar) bodies and *umwelten* has made our relationship a lateral one, an *Ineinander* that retains incomprehensible differences (Merleau-Ponty 2003, 271-273). Haraway describes the gendered projections of "mothering" that have shaped the selection and media characterization of women researchers like Goodall and perhaps tempted some to forget necessary restraints in their relations with primate subjects (Haraway 1989, 133-185). De Waal warns of the dangers of "Bambification"—sentimentalized anthropomorphism that infects many Walt Disney depictions of animals and often determines the plots of "wild life adventure" films (2001, 71-74). The lives of the chimpanzees Lucy, Nim Chimpsky, and Washoe, like that of Fern in *We Are All Completely Beside Ourselves*, and even Koko's and Kanzi's, have been violated and profoundly distorted by the problem of wanting too much, as if researchers sought to erase species barriers and fulfill a romantic dream almost straight out of *Tarzan*, but in the process forgetting to honor the integrity and dignity of the animal Other. This is a reversal of the situations in *The Epic of Gilgamesh* and *The Bacchae*, in which humans seek to deny their ties with the natural world. Here instead is what Hearne identifies as oppressive and insulting violation of another creature's privacy and independence (2000, 63).

A very different kind of primatology is practiced by my colleague Frances White, whose work in the Congo with wild bonobos is rigorously restrained to avoid the bonobos' habituation to humans. Building on traditions of careful observation of primates in their natural habitats begun by Jane Goodall, Kenji Imanishi, and other Japanese researchers as described above, it is related to field work of Dian Fossey and Barbara Smuts but conducted at a greater distance from the animals in order to prevent acculturation that would make the bonobos vulnerable to hunters of bush meat in the war-ravaged part of Central Africa that is their only habitat. White and her colleagues follow the bonobos they watch, but keep their distance on the ground while the bonobos go about their daily activities in the trees above. The bonobos recognize and tolerate their presence but sometimes play jokes by dropping fruit on them. Any kind of close relationship is avoided (Personal conversations and de Waal seminar March 2015).

## CONCLUSION

The modesty and restraint in this kind of attention to animals in their own worlds demonstrated by such primate research, and by many other ethologists and wildlife specialists observing animal communities more or less free of anthropogenic control, can prepare for a biosemiotically informed relational ethics. This would be based on something like the approach Val Plumwood described in opposition to the enormous arrogance she found in deep ecology: "On this relational account, respect for the other results neither from the containment of self nor from a transcendence of self, but is an *expression* of self in relationship, not egoistic self as merged with the other but self as embedded in a network of essential relationships with distinct others" (Plumwood 1996, 172). Philippe Descola describes the complex relational understandings of the Achuar people of Peruvian Amazonia, for whom most plants and animals have a soul similar to that of humans and live in an intersubjective ambience with them, characterized by mutual ethical responsibilities. All are persons within "theatres of a subtle sociability within which, day after day, humans engage in cajoling beings distinguishable from humans only by their different physical aspects and their lack of language." He finds comparable relationships with animals among the Northern Cree and Inuit of Canada, and in Mali and Sierra Leone, as well as in the Arctic and Southeast Asia (Descola 2013, 5, 14, 16, 27, 134). We cannot appropriate their world view; instead we must develop our own, shaped by what we now know about the biosphere, semiosphere, and overlapping worlds we share with all the others, and which, as Lestel reminds us, shape our imaginations and fullness of being in crucial ways (2013, 320). Derrida asks if we can imagine a world without animals (Derrida 2008, 7). The answer is no, for it would exclude us too; we are all part of each other; our *umwelten* overlap and inform each other in an intricately intercorporeal and biosemiotic reality.

As more and more parts of the earth become uninhabitable, we are likely to witness horrors never seen by our species or any other on such a scale. In *The Revenge of Gaia*, James Lovelock paints an extreme picture of the future on a rapidly warming Earth, predicting that "before this century is over, billions of us will die and the few breeding pairs of people that survive will be in the arctic region where the climate remains tolerable" (2006, xiv). We will lose our industrialized urban comfort zones and return to the way of life of our early ancestors. Who among our kindred organisms will accompany us? Many times in human history complex societies have collapsed; we have only to remember the Mesopotamians, Minoans, and Mayans among others. Nature will take back our constructions very quickly, as we have seen in Chernobyl, and as Alan Weisman predicts in *The World without Us*, but with a difference caused by increasing heat. In a new essay about the recent migration of the Golden Jackal into Estonia from Asia Minor, Timo Maran analyzes the complex negotiations and adaptations that have to occur as species encounter changing environments and new relationships (Maran, 2015). In fact, most of human history has been lived in constantly changing communities over geological time, where people have had to take care to respect the power and distinctive being of a myriad of other animals sharing their landscapes, water, and food sources. Such a world is represented in the cave paintings of southern Europe and the rock art of the Mongolian Altai, as well as in many other places where our ancestors left the traces of their imaginative understanding of their world. The ecological imaginary we inherited from them must be reawakened and adapted for our own changing circumstances. Descola cautions that we must find a way to respond to "the challenge of recomposing into viable and unified groups an ever-increasing number of existing beings needing to be represented and treated equitably. It is up to each one of us, wherever we may be, to invent and encourage modes of conciliation and types of pressure capable of leading to a new universality that is both open to all the world's components and also respectful of certain of their idiosyncrasies" (Descola 2013, 405-406). We must find our way to a renewed understanding of these relationships through the radically complex knowledge of shared bodies and dynamic communications within "an ensemble of overlapping horizons, lifeworlds, and styles of being" (Singer 2015, 114) that evolutionary biology, genomic studies, and biosemiotics have opened up for us.

This must be very carefully done to maintain rigorous ethical questioning of where it takes us, so that we embrace a relational view of ourselves within the semiosphere and develop an appropriate communal sense of ethical behavior that moves far beyond the narrow western philosophical discourses of "rights," ability to suffer, cognitive hierarchies, and utilitarian valuation of life forms. As Morten Tønnessen and Jonathan Beever explain, "The overlapping *Umwelten* in and through which all living organisms exist constitute the vast web of ecological relations. Understanding the nature and value of those webs of relations from a biosemiotic perspective will move environmental ethics beyond sentience" (2015, 58). We will have to relearn how to live together by recovering in the imaginary the possibilities for restraining our manipulations of the landscapes and seas, while opening ourselves with proper attentiveness to be taught by our animal Others.

# THREE

# Animals in a Noisy World

*Almo Farina*

Acoustic noise is a widespread phenomenon in nature, which has been hampered in the current Anthropocene epoch by recent growing human intrusion in many ecosystems. Noise can be defined as sound characterized by poor information (high level of vibrational disorder) that masks other sounds and that affects the active space used by terrestrial and aquatic animals for acoustic communication. From a human perspective, noise is defined as an unwanted/unpleasant sound and is considered a type of environmental pollution.

Noise sources may be generated by natural (geophonies: e.g., wind or heavy rain; biophonies: e.g., animal choruses) or by man-made processes (technophonies: e.g., combustion engines or gear friction). In the last case, transportation facilities and industries are major sources of acoustic intrusion that represents, especially in urban areas, the second major source of social concern after the poor quality of air. The exposure of people to long periods of noise has been proven responsible for different levels of annoyance and other negative effects (e.g., in communication, recreation, concentration, and sleeping activities).

Acoustic noise impacts important functions in animals, such as habitat selection, pair formation, resource tracking, and prey-predator mechanisms. Additionally, a differentiated species-specific tolerance results in an important factor responsible for significant changes in community composition. To reduce such negative effects, adaptive mechanisms like the increase of the amplitude of acoustic signals (Lombard effect) or the shift of signal frequency have been observed in many acoustics animals.

If noise represents an important environmental cause of degradation and source of stress for several organisms, especially in rural and urban (metropolitan) areas, underwater noise in the marine system is one of the major environmental threats to marine fauna because noise is not only confined to developed coasts but is largely dispersed in oceans along trading routes. In addition, underwater noise, due to the speed of acoustic waves in liquids, has large influences in several marine systems affecting mammals and pelagic fish, which poses true challenges in environmental mitigation, remediation, and ecosystem conservation. In addition, oil spills and other seismic geological prospections when carried out in remote territories and in oceans are further important sources of noise affecting animal populations, producing changes in their behavior, and having consequences in the trophic webs.

In this paper, the ecological consequences of acoustic noise are discussed in light of recent theoretical principles and methodologies of ecoacoustics and soundscape ecology with the hope to contribute to the solution of the problems created by acoustic pollution both in human populated areas and in remote and fragile areas.

## INTRODUCTION

Nature is rich in sounds that are the product of geophysical and biophysical dynamics. In this loud world, acoustic animals play an important role, like components of a "Great Animal Orchestra" (Krause 2012). However, during the last two centuries considered by the Nobel prize laureate Paul Crutzen and his colleague Eugene Stoerner, a new geological epoch (the Anthropocene) (Crutzen and Stoermer 2000), these sounds have been masked or substituted by new sounds of anthropogenic origin (Pivato 2011).

The recently established field of ecoacoustics (Sueur and Farina, 2015) has focused the attention of researchers and scholars on the ecological role of sound in natural and anthropogenic dynamics. To better understand the complex interaction between species and their sonic context, the concept of sonic context or soundscape as a spatial representation of various sources of sounds (Pijanowski et al. 2011), which is associated with the concept of acoustic community, has been extensively utilized.

The word *noise* derives from the Latin *nausea* and can be defined as an unintentional background sound that masks acoustic signals. The Canadian composer and scholar Barry Truax proposed several definitions of noise (e.g., unwanted sound, unmusical sound, or any loud sound or disturbance in any communication system) (Truax 1999). According to the signal detection theory (SDT) (Green and Swets 1966), "background noise" may be defined as any energy in the environment that is irrelevant to the communication between a signaler and a particular receiver (Luther and Gentry 2013).

Generally, noise is a sound characterized by high amplitude like thunder or the sound produced by waterfalls. However, noise can be also defined as a sound that has no explicit information, like the buzz of a multitude of people. Noise, as every intruding sound, has no boundaries, it is difficult to control and manage, and when massively present, as in urban areas and inside or around great logistic hubs (harbors, airports, and railway stations) or linear transportation infrastructures (railways, aerial corridors, highway, and shipping lanes), it becomes a source of chronic disturbance with severe consequences to human health and beyond (Babisch et al. 2005; Barber et al. 2009).

Therefore, according to a social or psychological approach noise is defined as every sound whose characteristics produce annoyance or health consequences in exposed people. In fact, the effects of noise pollution on humans are numerous, pervasive, and persistent as well as medically and socially relevant as recently argued by Goines and Hagler (2007), and a cumulative effect on young generations is expected (van Kempen et al. 2010). In 1971, the World Health Organization (WHO) announced that noise is a major threat to human well-being.

According to a semiotic perspective, noise is considered to be every sound that has small intrinsic information. According to an ecoacoustic approach (Sueur and Farina, 2015), noise pertains to the portion of an acoustic spectrum that does not offer information per se; nevertheless, noise and its distribution in time and space can be used to detect events and to assess the acoustic quality of locations across a landscape with important economic consequences.

The soundscape is heterogeneous, and this means that there is an alternation of quiet and noisy acoustic patches or sonotopes (Farina 2014). Sonically tolerant species can establish in both conditions and can cope with local sonic conditions, but narrowly sonically tolerant species can select only quiet areas. Unfortunately, in the modern world dominated by humans, the heterogeneity of the sonotopes is decreasing, favoring permanently homogeneous noisy sonotopes that can become hostile to many forms of animals.

A noisy world is associated with human (technological) intrusion in the ecosystems, and it returns the image of an "aggressive" source of environmental pollution that alters the human perception of the natural world and poses a serious challenge to several delicate processes related to behavior and physiology of animal organisms in terrestrial and in aquatic systems (McDonald et al. 2006; Watts et al. 2007). For this reason, the noise generated by human activities represents a phenomenon that, with the countless implications on human wellbeing and with the contemporary growing impact on natural systems, has inspired an impressive number of research studies conducted by biologists, ecologists, and resource managers and has filled the agenda of policymakers (Luther and Gentry 2013; Farina 2014) with relevant implications for ecological conservation (Slabekoorn and Ripmeester 2008; Brumm 2010). Effects of noise have been considered also on domestic (farming) animals. Noise is high in modern farms and when animals are transported by trucks or airplane cargos and ship. The effects are similar to the ones found in humans: increased hormonal production, increased heart rate, and reduction of production (e.g., Broucek 2014), although domestic animals have a higher sensitivity to higher frequencies than humans (Phillips 2009).

Based on the origin, noises may be generated by natural sources (geophonies: e.g., strong wind, volcano eruptions, earthquakes, thunder, and running water; biophonies: e.g., breeding or roosting of colonial birds and mammals), or by anthropogenic sources (technophonies: e.g., urban traffic, oil spills, and stationary machines like air conditioners). Noise may be the emergent pattern produced by the sum of different sounds that interfere with each other, such as the sounds produced by several people speaking in a small space (e.g., inside an overcrowded bar). Usually noise is the result of exceeding amplitude and the duration of one of these three components that compose an acoustic patch or sonotope (Farina 2014, p. 18–19).

In natural environments, noise is a less frequent event than in human transformed ecosystems where the main source of noise is the result of technological machinery; however, in some places, such as near an active volcano or along sea coasts, a diffuse and persistent amount of (natural) noise can be present. In tropical forests, dawn and dusk choruses of insects, frogs, and birds often are so loud as to create a noisy effect that prevents the distinction of individual signalers.

Today, detailed and well-documented knowledge of the effect of large and prolonged doses of noise on human behavior and health are available (Goines and Hagler 2007). On the other hand, much empirical evidence has been collected on the effects of the noise on wild animals. However, how acoustic animals perceive such types of human-categorized sound is only partially and imprecisely known. Furthermore, few studies have been conducted on the physiological effects of noise on animals, especially long term, but it is likely that noise may have effects on animals that are similar to the effects on humans.

## NATURAL SOURCES OF BACKGROUND NOISE

Natural sources of background noise can be categorized into abiotic and biotic origins (table 3.1).

### *Abiotic Sources*

Abiotic sources include rivers and waterfalls, which may have a peak of frequencies below 1 kHz and in some cases, may be above this. Every

**Table 3.1. Commonest sources of noise**

| | |
|---|---|
| Natural | abiotic [river, waterfall, wind, rain, eruptions] |
| | biotic [wild animal choruses] |
| Anthropogenic | traffic [airplane, car, train, boat], industry, construction, urban |

species that vocalize around these frequencies have problems when transmitting signals. Wind and air turbulence produce problems in transmission of 200 Hz; rarely does wind interfere with biotic signals above 2 kHz. Wind regime, absent in early morning and more frequent with breezes in the late morning and at dusk, has a strong influence on signal transmission. Wind acts differently at different heights in the forest. For instance, in tropical forests, the wind effect is clearer in the mid-story and canopy. Rain, when intense, may produce sounds with frequencies of up to 5 kHz, interfering with the majority of bird songs. In general, rain and wind depress vocal activity in many species (Robbins 1981; Bruni et al. 2014).

The adaptation to abiotic noise, especially river and waterfall noises that are spatially located and temporally constant, has been proven in different species of frogs and birds. Evident adaptation is present in species that live in windy regions like the Western Mediterranean that have songs and calls that are redundant. In particular, grassland birds have signals at high redundancy to be detected and discriminated by receivers (Brown and Handford 1996).

### *Biotic Sources*

Insects (orthopterans and cicadas), operating at a signal of up to 4 kHz, are more widespread sources of biotic noise in tropical regions. Their activity reaches the highest intensities during dusk with frequencies between 3.15 and 12.5 kHz. Some timing has been observed in tropical Borneo with a temporal partitioning between insects (cicadas and crickets) and frogs. However, there are differences according to regions and seasons. Acoustic interferences are expected for colonial vertebrates, and this represents a true challenge in discriminating individual signals; for instance, *Hyla microcephala,* during its chorus, inhibits the song of the male of *H. ebraccata* (Schwartz and Wells 1983).

## ANTHROPOGENIC SOURCES OF NOISE

Anthropogenic noise (technophonies) is the result of urban life, transportation, industry, and energy production and is increasing with the increase of human intrusion. Noise arises from highways, railways, harbors, and airports. Transportation noise has a cylindrical (geometric) spread with a decline of 3 dB, which doubles the distance when compared to a point source that has a decline of 6 dB, doubling the distance. This is expected to have a great impact on the environment in the future. Noise produced by a car at 80 km/h and at a 3-m distance produces a noise of 71 dB, but a truck in the same conditions has sound levels of 85 dB. The increase of truck traffic is expected to have a great impact on the environment. The effects of traffic noise are affected by the time at which the rush occurs. Generally, peak traffic occurs in early morning and at dusk, overlapping with the greater acoustic activity of organisms, and this creates an important impact. Trains are very noisy (80 dB) but are more intermittent than car traffic. Gas compression stations (Bunkley et al. 2015) and surface mines (Duarte et al. 2015) are important areas of noise pollution. In water, anthropogenic noise is transmitted with higher efficiency, in particular the noise produced by boats and ships. Seppänen and Nieminen (2004) measured such noise at 50–100 meters from the waterways, recording a noise of 120–140 dB (linear SPL, re 1Pa) at prevalent frequencies ranging from 1 to 5 kHz.

## EFFECTS OF NOISE ON THE ANIMAL WORLD

During the last decennium, the relationship between noise and animals has received a lot of attention in animal behavior studies (c.f. Klump 1996) and only recently has been the object of investigation in ecoacoustics (e.g., Slabbekoorn et al. 2010) (table 3.2). Among vertebrates, more studies on noise impact regarding birds have been carried out. Some of the effects of noise pollution on this group of vertebrates include physical damage to ears, stress responses, fight-flight responses, avoidance responses, change in other behavioral responses, such as foraging, changes in reproductive success, changes in vocal communication, interference with the ability to hear predators and other important sounds, and potential changes in populations (Ortega 2012).

The ecological and ethological role of noise remains highly disputed; in fact, noise is considered in animal communication studies as "any

**Table 3.2. Principal masking effects of noise on animals**

| | |
|---|---|
| Reduction of the active space | |
| Changing communication | disruption of social aggregates, alteration of the prey-predator relationship |
| Decrease of the transmission distance | |
| Physiological stresses | |

factor that reduces the ability of receiver to detect a signal or to discriminate one signal from another" (Brumm and Slabbekoorn 2005), but according to a ecosemiotics perspective, noise per se is a source of information. For instance, heavy rains or strong winds become the context in which organisms interact, and it represents a typology of ecological code (Farina and Pieretti 2014).

### *The Active Space*

The active space is defined as the distance at which a signal can be detected and decoded (Marten and Marler 1977; Brenowitz 1982; Klump 1996). Outside the active space, the sound perceived does not have enough information to be properly decoded and transformed into meaningful information. This distance is the limit of the sensory threshold. The physical attenuation of a signal is not the only factor that delimits the active space; noise may have a strong influence on this process. Noise reduces the ability of a receiver to discriminate the signal from the background noise. For instance, the active space of *Strix aluco* can be reduced 69-fold (from 118 ha to 1.7 ha) during heavy rain (Lengagne and Slater 2002). Investigation of the brown-headed cowbird (*Molothrus ater*) by Gall et al. (2012) proved that the active space is affected by both habitat type and level of urbanization.

Animal physiology and behavior are affected by noise in different ways producing annoyance, chronic stress, and hearing loss, and usually a noise environment is avoided by less tolerant species.

Animal communication, if masked by anthropogenic noise, may have negative effects at the individual, population, and community levels (Francis et al. 2009). Individual fitness is expected to be depressed in organisms living in a noisy environment because this means using more vocal energy to assure the communication level. In many cases, when sound is used to mate, such patterns can result in degradation, violating the theory of honest signaling (Zahavi 1977).

Background noise, by masking the quality of the signal (e.g., the song quality of males), reduces or suppresses key signals, such as an alarm call, thus increasing the predation risk. Negative effects are evident in the relationship between the prey-predator signaling systems, altering the signals necessary to drive this complex relationship. As outlined by Slabbekoorn and Ripmeester (2008), noise produces direct stress, masks the arrival of predators or the associated alarm calls, and generally interferes with communication mechanisms.

The effect of noise can be evident in social animals, producing great difficulties to maintain social aggregation or preventing the exchange of strategic information like the location of areas with abundant food. Dawn and dusk bird choruses (Farina et al. 2015) and long-distance signals (e.g., in African elephants) (McComb et al. 2003) are used to inform individuals about reciprocal presence. To be effective, these signals require low background noise.

Animal communication is based on the signal-to-noise ratio, and this ratio depends on the amount of background noise that is present in an environment at a specific time of day or during a season. For instance, background noise and bird acoustic patterns are strictly related, as emphasized by Slabbekoorn (2004). Several animals use acoustic signals to defend territories and to collect other information to track the necessary resources to maintain individual fitness. Finally, living in a habitat with a low level of background noise is essential to attract mates.

The effects of noise can be proved concerning potential fitness costs, as argued by Read et al. (2013). Background noise may decrease the transmission distances according to the "acoustic adaptation hypothesis" (Morton 1975). When a signal is shifted by a species due to noisy background toward a higher frequency, this signal can be easily attenuated when compared to a low-frequency signal. This shift produces a reduced acoustic range on individual fitness. Similarly, the alarm call of many passerines is uttered at higher frequencies associated with an increase of amplitude to maintain efficiency, and this exposes birds to a higher risk of predation. In some cases, to avoid the masking effects of noises, birds move to higher perches, and this increases the exposure to higher predation risk.

Physiological stresses have been proven in female wood frogs (*Lithobates sylvaticus)* caused by traffic noise, according to Tennessen et al. (2014). Individuals were found with an increase value of glucocorticoid hormone (plasma corticosterone) that may have deleterious consequences on survival and on reproductive capacity in this species. Noise may be a further cause of frog decline due to their sensitivity to noisy environments (Bee and Swanson 2007). In Yellowstone National Park, the intrusion of snowmobiles has been proven by Creel et al. (2002) to increase the fecal glucorticoides (GC) of elk (*Cervus elaphus*).

In many species, mating success is linked to communication, and the increase in background noise may have a strong effect on species fitness. We do not have much evidence of the effects of different types of anthropogenic noise. Noise is a sonic pattern that can be considered to have a negative impact for some interacting species. Conversely, it may have beneficial results for others, if they are in competition. Francis et al. (2009) investigated the effects of noise on the prey-predator relationship in a community of birds living in urban areas. Noise seems to alter the relationships between species when some species (predators) avoid noisy areas that are occupied by prey that are noise-tolerant species. These latter species find a successful habitat in urban noisy areas following a disruption of species interactions, which can determine the homogenization of avian communities.

Some species of bats may be strongly affected by traffic noise, resulting in a decrease of the rate of prey captured. In a noisy condition, such as near a gas compressor station, Bunkley et al. (2015) found a 40 percent reduction of echolocation search call activity of Brazilian free-tailed bats (*Tadarida brasiliensis*) compared to quieter locations. This species has the capacity to change the length of its call sequence as a result. These authors have also found that some species with a call frequency >35 kHz exhibited an acoustic activity not affected by noise sources. However, species with an acoustic activity <35 kHz were affected by noise.

## STRATEGIES TO NEUTRALIZE THE EFFECTS OF NOISE

The consequences of a noisy background on animals appear to be a complex issue because the effects vary among species and individuals as well as the time of year and day. Different strategies may be used to reduce such effects, such as changing amplitude, frequency, time of vocal activity, signal redundancy, etc. (table 3.3).

### *Changing Amplitude*

The amplitude shift has been described for the first time in humans by the French otolaryngologist Lombard Etienne in 1911, since then known as the Lombard effect. The Lombard effect is the tendency for a human speaker to increase the intensity of vocal level to improve the intelligibility of a speech signal in the presence of a loud background noise (Lombard 1911). After more than 60 years, Potash (1972) observed that the Japanese quail (*Coturnix japonica*) changes the amplitude of their vocalizations when exposed to loud white noise. More recently, Brumm and Todt (2003) proved the Lombard effect on an urban population of the nightingale (*Luscinia megarhynchos*). This species was able to increase the amplitude of their sound during working days and decrease the amplitude during the weekend. In addition, experiments conducted by Cynx et al. (1998) on zebra finch (*Taeniopygia guttata*) have proven that individuals, when exposed to different levels of noise, were able to change the amplitude of their vocalizations. In a marine environment, Holt et al.

**Table 3.3. Some strategies to reduce the effect of noise**

| |
|---|
| Change of signal amplitude |
| Change of signal frequency |
| Change of signal redundancy |
| Change of time of acoustic activity |
| Change of behavior |

(2008) described an increase of the call source level by 1 dB for every 1 dB increase in the background noise level in killer whales (*Orcinus orca*). The California ground squirrel (*Spermophilus beecheyi*), a mammal that lives in human-modified grassland landscapes, under different levels of noise masking (Rabin et al. 2003), has a broad repertoire of chatters, whistles, and single-note repetitive calls according to the presence of potential predators. In the presence of highway traffic noise, this species has the capacity to shift acoustic energy in those harmonics that do not overlap with highway noise. This species seems to have a high acoustic plasticity. The energetic budget necessary to increase the amplitude may be reflected in negative effects on the metabolism and on the energetic budget itself.

*Changing Frequency*

The change of frequency of vocalizations is another strategy to reduce the masking effect of background noise. This effect has been described in the concave-eared torrent frog (*Amolops tormotus*), a species of arboreal frog of Central China that uses ultrasonic harmonics to communicate in a noisy environment (Narins et al. 2004). A similar effect has been observed in the black-faced warbler (*Abroscopus albogularis*), an oscine songbird that lives close to the torrents in a constantly noisy environment and that uses ultrasonic frequencies to communicate (Narins et al. 2004). These two cases may be considered examples of the effect of a long-term co-adaptation to a noisy environment. In addition, a similar effect has been described by Slabekoorn and Peet (2003) in great tit (*Parus major*) populations. Urban populations living in noisy conditions were observed to shift the song repertoire toward higher frequency bins.

Mockford and Marshall (2009), through the use of playbacks, have demonstrated a difference in spectral aspects of the song of the great tit (*Parus major*) living in urban and rural environments. Males living in quiet areas had significantly stronger responses (e.g., they were faster to sing over the playback song and approached the speaker) when compared to males living in urban areas. Rural populations have lower minimum frequencies than individuals from urban areas. Changes in frequency have also been proven in three dialects of the white-rowed sparrow *Zonotrichia leucophrys* living in the San Francisco peninsula (California) with a distinct frequency range along urbanization gradient (Luther and Baptista 2010).

In roadside habitats characterized by vehicle traffic, Lampe et al. (2012) demonstrated that the grasshopper *Chorthippus biguttulus* has the capacity to change the frequency of the song by elevating the frequency maximum. In a later study, the same group of researchers (Lampe et al. 2014) demonstrated that this species is able to develop plasticity in mating sound when nymphs are exposed to different levels of background sound. Males exposed to road noise during the nymph period developed signals at higher frequencies compared to males exposed to quiet conditions. However, higher frequency signals and an increased syllable-to-pause ratio that represents a sexually selected signal trait were found in males living in roadside areas.

There is experimental evidence that birds with lower frequencies are more affected by urban anthropogenic noise and that the minimum song frequency is highly predictive of species' responses to urban noise when other environmental characters like vegetation are excluded. Many species of birds can adopt habitats in green spaces (parks, plantations, recreation areas, etc.), but the level of acoustic pollution can make a difference. For instance, in Spain and Portugal, Paton et al. (2012) found 91 species of birds, but at least 10 (*Regulus regulus, Streptopelia turtur, Dendrocopos minor, Buteo buteo, Hirundo daurica, Corvus corax, Oriolus oriolus, Cettia cetti, Passer hispaniolensis,* and *Sylvia melanocephala*) seem affected by the noise level. A reduction of noise level could encourage the settlement of rare species, reducing the risk of bird homogenization (Joo et al. 2011).

### *Changing Signal Redundancy*

According to information theory, signal redundancy is a mechanism that is used to improve information transfer in the presence of noise (Wiley 1994). Signal redundancy, represented by an increased number of syllables, has been demonstrated in Japanese quail (*Coturnix japonica*) by Potash (1972) when animals were subjected to an increased level of background noise. The same effect has been observed by Diaz et al. (2011) in European serin (*Serinus serinus*), which sang more in the presence of traffic noise during working days, and it shows reduced singing activity during the weekend when traffic is low. Similarly, singing males of chaffinch (*Fringilla coelebs*) were observed to extend the length of their signals with more notes in the presence of environmental acoustic noise (i.e., close to waterfalls and torrents) (Brumm and Slater 2006).

### *Changing Song Timing*

A further technique to reduce interference between species is separation in time. For instance, Luther (2009) found more differences in the acoustic community of a vertical stratum of a tropical forest than between the different strata. Species living in the same forest stratum presented the most dispersed song. In fact, the song of congeneric species were more dispersed than songs of random species. In the coqui tree frog (*Eleutherodactylus coqui*), Zelick and Narins (1985) demonstrated that individual males were starting with a delay of 750 ms when exposed to playback noise.

*Changing behavior*

High amplitude and constant noise, such as those in areas close to rivers or those of insect choruses, favor adaptation, while episodic noise may produce plastic adjustments, such as the alteration of signaling or the selection of notes or syllables in order to increase the signal-to-noise ratio. The plastic adjustments are considered a change in behavior. For instance, road traffic noise has been shown to affect the prairie dog *Cynomys ludovicianus* by lowering aboveground activity, reducing foraging, and increasing the vigilance according to the risk disturbance hypothesis (Shannon et al. 2014).

Wind turbines are spreading in rural and forested landscapes, resulting in a source of noise. For instance, Rabin et al. (2006) observed complex interactions in a population of California ground squirrels (*Spermophilus beecheyi*) at the Altamont Pass Wind Resource Area of Northern California. The alarm call of this species, which is important to avoid predation, is masked by wind turbine noise. The authors selected two populations: a control group far from the wind farm and an experimental group close to the wind farm. The population living close to the turbine was noted to demonstrate new behavior consisting of an elevated rate of vigilance and a quicker return to their burrows when alarm calls were uttered.

Experimental evidence has demonstrated that, in European robins (*Erithacus rubecula*), a high noise level can produce changes in behavior and cause males to move away from the source of noise. At the same time, it has been shown that males increased the frequency and reduced the complexity and duration of a song (McLaughlin and Kunc 2012). Changes in behavior, such as in tonic immobility, were observed in toads (*Bufo bufo*) (Lupo et al. 1991). In shore crabs (*Carcinus maenas),* Wale et al. (2013) experimentally found that the playback of ship noise produced changes in crab behavior, including disrupting feeding and slowing the reply to retreat to shelter.

In field crickets (*Gryllus bimaculatus),* Schmidt et al. (2014) experimentally proved that the females have a weaker reply to male songs when background noise is present, despite the fact that the song of this species is distinct (no frequency overlap) from the background noise. An explanation may be that females that should have a broader auditory tuning are more susceptible to anthropogenic noise that produces a distraction.

## NOISE IN MARINE AND FRESHWATER SYSTEMS

A distinct section for considering noise in marine systems is required. This is largely due to the great importance that noise pollution has on the life of aquatic systems and the extent of this type of anthropogenic pollution, which is strictly connected with the economic value of global trading (Frisk 2012). In an aquatic environment, sounds are transmitted approximately five times faster (1484 m/sec) than in the atmosphere (343 m/sec). In addition, the distance at which a sound travels with negligible degradation due to the lack of obstacles is longer than in terrestrial systems.

Due to an increase of trade and touristic shipping, the ocean soundscape is undergoing rapid changes, which produce a dramatic impact on animal life and its communication processes. Chronic forms of human-generated ocean noise have the same effects as the loss of environment in terrestrial systems. Further sources of noise intrusion are represented by the location of wind farms in shallow waters (Wahlberg and Westerberg 2005) and by the seismic exploration of the deep sea, especially at higher latitudes encouraged by ice reduction (Klinck et al. 2012). Recreational motorboats have had a growing impact on the fish community not only because of the physical disturbance and chemical pollution but also because of the noise produced by the engines, as recently argued by Whitefied and Becker (2014). A further source of acoustic impact of anthropogenic noise is represented by the emission of medium and high frequencies, mainly by warships (Koper and Plön 2012). These frequencies have an impact on whales and dolphins and in many cases seems to have caused several cetacean strandings.

The ocean acidification by carbonic acid has modified the sonic context; acoustic waves are less absorbed by acid water and sound propagates farther. As a result, an increase in the background noise of oceans, which has been 2.5-3 dB per decade in the frequency band of 30-50 Hz, has been reported (e.g., Etter 2012; McDonald et al. 2006; Simpson et al. 2011). This is the result of previsions based on geometric models. Experiments have demonstrated that the cumulative effect of anthropogenic noise may be underestimated by applying geometric models in coastal shallow waters (Pine et al. 2013), and new models are necessary to better predict the anthropogenic noise.

Aquatic life, including crustaceans, fish, and mammals is substantially affected by noise. The importance of the sonic environment in the marine system is demonstrated by the behavior of coral reef larvae, which have been proven to utilize background noise of fish and crustaceans of the coral reef as a phonotaxis source and to select settlement locations during their planktonic life (Tolimieri et al. 2000). In this case, the importance of noise mitigation of anthropogenic noise along the coral reef appears extremely important for the coral recruitments, confirming that the soundscape has an "immense value to the pelagic larval stage of a coastal organism" (Vermeij et al. 2010).

The impact of noise on fish (especially in marine systems) is just the beginning of the story. In fact, there are more challenges to evaluate the impact of anthropogenic noise on the lives of the estimated 32,000 species of fish, as recently argued by Radford et al. (2014) and on which we have little information concerning their acoustic activity. The fragility of fish to noise was proven by McCauley et al. (2003) in experiments conducted on pink snapper (*Pagrus auratus*) where the hair cells were damaged after exposure to air-gun noise.

The ambient noise may have an impact on predatory risk. This effect has been illustrated by Chan et al. (2010) in the Caribbean hermit crabs (*Coenobita clypeatus*), which showed a reduction in attention resulting in easily being approached when exposed to boat noise playback.

Pile-driving activity is very common along coasts and offshore. This source of noise affects many pelagic mammals, like bottlenose dolphins, that are damaged when within 100 m from the pile-driving, and further behavioral modifications are expected at a distance of 50 km (Bailey et al. 2010).

Au and Green (2000) demonstrated that boats close to pods of humpback whales (*Megaptera novaeangliae*) could create communication problems in these animals during the singing season (February-April) in Hawaiian areas. In fact, this species utters songs between 170 and 175 dB (re 1Pa at 1 m) that are lower than the noise produced by some boats (120 dB at 91 m). In right whales (*Eubalena glacialis* and *E. australis*) that are exposed to an increased level of ambient noise, an increase of ambient noise level produces a lower rate call, probably to avoid the masking effect from low-frequency noise (Parks et al. 2007).

Changes in behavior have also been observed in manatee (*Trichechus manatus latirostris*) by Miksis-Olds and Wagner (2011). When ambient noise is high, this species spends more time feeding and less time milling (indirect movement). This is probably a direct consequence of the hostile environment created by high noise that forces this species to satisfy nutritional requirements in a shorter time to escape a potentially dangerous situation.

Protecting hot spots for marine mammals is not an easy task. Williams et al. (2013) measured the level of ocean noise in Canada's Pacific Ocean for fin, humpback, and killer whales. These authors found no correlation between protected areas and noise levels. This evidence requires a better use of field information to formulate effective policies.

## CONCLUSION

The human population is growing very quickly (Cohen 2003) and is concentrated in urban areas (54 percent in 2014, WHO), which represents a strategic environment for humans and for many other organisms. Noise is an impacting phenomenon in such areas, and scientists and policymakers will be confronted with this subject in the future because the consequences for people and animal species are not any less important than other types of anthropogenic pollution.

It appears clear that several actions used with an ethical perspective to improve human well-being, like the use of renewable resources or the creation of therapeutic landscapes to recover from the stress of human lifestyles (Farina et al. 2007), unintentionally result in benefits for a large set of "neglected" organisms, and this fact may represent an important guidance for assuring a reasonable future for biodiversity.

Quiet areas, a value largely neglected in the past, are now considered an important character of an ecosystem. Living in a quiet world is, for modern societies, a value equivalent to having access to quality food, drinkable water, and clean air (Kaplan and Kaplan 1989). Moreover, the quality of the sonic environment has been demonstrated to be important for young people living in urban landscapes (Hedblom et al. 2014). In addition, in order to have recreational time far from noise, such as that found in national parks, people pay extra to have better noise regulation (Merchan et al. 2014).

Noise mitigation could improve the habitat suitability of urban areas for species that include low-frequency elements in their songs (Proppe et al. 2013), producing an increase of biodiversity. Nonetheless, mitigation strategies remain controversial. This is the case in the mitigation of the road impact in Australia by the installation of engineered structures (Goosem 2012).

The concentration of millions of people in accessible front-country areas creates problems concerning the natural quiet, with no easy solutions for local development and tourist expectations, as outlined by Cessford (2000) for remote trails in New Zealand. The soundscape is considered an important component in heritage conservation (O'Connor 2008). Recently, many efforts have been dedicated to protecting national parks and other protected areas from anthropogenic noise (e.g., Pilcher et al. 2009).

One of the most "spectacular" effects of human intrusion on a planetary scale is represented by climate changes. A recent Intergovernmental Panel on Climate Change report recognizes that more than 50 percent of the increase of surface temperature is a result of direct human activity (IPCC 2013). The increase of temperature is expected to affect the phenology and physiology of organisms, range and distribution of species, distribution of plants and animals, their displacements, recombination of communities, and the changes of the structure and dynamics of ecosystems (Walther et al. 2002). In this scenario, the acoustic mechanisms of vocal animals may be strongly disturbed. In particular, acoustic communities may change composition and dynamics, and this could open new perspectives regarding the effect of anthropogenic noise.

The acidification of oceans will increase sound transmission with negative effects on several marine organisms that will be more exposed to anthropogenic noises. However, other threats, like wind farms along coasts and on mountain ridges, new airports, spreading of gas compressors in remote areas (e.g., Siberia, Alaska) or urban sprawls are further sources of anthropogenic noise that will create a more hostile environment for several animals.

Ecoacoustic monitoring could facilitate our understanding of the new dynamics introduced by the effects of climatic change, as suggested by Kloepper and Simmons (2014), and help assess the level of the growing anthropogenic manipulation of marine and terrestrial soundscapes in a world in which the energy demand is expected to grow.

# *II*

# Phenomenology in the Anthropocene

# FOUR

# A Phenomenological Approach to the Imaginary of Animals

*Annabelle Dufourcq*

The starting point of my reflection[1] was the following problem: There is a striking contrast between, on the one hand, the imaginary (i.e., the imaginative tradition) of talking and metamorphic animals in art, literature, myths, and popular imageries and, on the other hand, the classical rationalist representation of animals as silent and cooped up in predictable patterns of behavior. The first representation is commonly regarded with suspicion and often discarded as a mere fiction, the result of an illegitimate anthropomorphic projection, while the second is generally considered as realistic. Nevertheless, contemporary researches in ethology, biology, and biosemiotics, as well as a critical philosophical approach have shown that the first rationalist model is a construction, somehow a fiction (Derrida 2006, 2008, 2010), while, correlatively, the consistent fascination exerted by the first representation and its status of transcultural archetype still requires to be explained. The close similarity between animals' behavior and human dreams has been highlighted by several thinkers (Bateson 1972, 60–69; Maran 2012; Portmann 1965, 240; Ruyer 1953, 853; Merleau-Ponty 1995, 246), but we still have to discard the suspicion of anthropomorphism that could be cast on such a claim.

Imagination obviously has access to a dimension that escapes a rationalist approach. But I would like to examine the precise characteristics of this "dimension" and if, through imagination, maybe in a privileged way, a genuine *knowledge* of the very nature of animals themselves can be reached.

To that purpose, I would like to show that a phenomenological approach to imagination and the imaginary constitutes a decisive means of escaping a traditional definition of imagination as a mental faculty that only composes subjective and arbitrary representations. Phenomenology—I will refer more specifically to Husserl's and Merleau-Ponty's analyses—allows us to understand imagination as entering into relation with the very being of the imagined objects, a being that lends itself to a discovery *through imagination,* because it possesses specific characteristics (ambiguity, creativity, ubiquity, expressiveness) that must remain inaccessible to reason and to any objectifying approach. On this basis it becomes possible to assert the value of a discovery of animals through the imaginary, an approach without which animals' ability to enter into relation with the dimension of open meanings, interpretation, and creation of forms would remain beyond reach. Only creation may understand creation, in a new sense of the concept of knowledge. As a result, the imaginary does not simply give access to an archaic or simply fuzzier, less developed, or less sophisticated form of meaning than the ones that may be found in a rationalist and logical type of thought: it is the only way to acknowledge and encounter a subjectivity. I would like to claim that the "human" imaginary is an interface that gives access to the other animals' imaginary. Imagination—as the faculty that enters into dialogue with the imaginary as an interspecific semiotic interface—is the source of a new form of knowledge essentially connected to ethical concerns.

## TRANSPOSITION, METAPHORS, AND ARCHAIC FORMS OF MEANING

The essential link between the knowledge of animals and imagination appears under various forms. I would like to present three of them, among the most recurring and decisive, so as to better define the problems at stake. My contention will be that these problems cannot be solved as long as we remain in the framework of a classical definition of imagination as the faculty that produces arbitrary associations and fuzzy meanings.

### *Empathy*

The question of the imaginary of animals can first be approached through a problem that has now become a touchstone in philosophical reflections on animality: is it possible to transpose myself into other animals? According to Nagel's now classical question: may I know what it is like to be a bat?

As put by Elizabeth Costello in Coetzee's *The Lives of Animals*: we definitely can imagine what it is like to be a bat, human imagination is actually extremely fertile in this regard,[2] but another important question is the following: is empathy a way to *know* animals? May I use it as a rigorous method that would give access to a genuine knowledge of other animals' very being?

To be sure, there is a form of withdrawal of animals, of silence and an abysmal mystery that lead von Uexküll (1934) to speak of separate worlds and which, for instance, appears in a particularly tragic way in Herzog's *Grizzly Man:* Timothy Treadwell's imagery of bears and his immoderate resort to empathy lead him to a deep misunderstanding of their behavior. Timothy kept speaking to the grizzlies, he was living beside them, but it remains doubtful that he ever set up any dialogue with them, nor was integrated rather than tolerated within their territory, nor built any common *world* (more simply, strictly speaking: a world) with them.

As a result, one may rightly be tempted to claim, following Heidegger (1983), that our transposition into animals cannot be achieved, while it is an immediate and obvious aspect of the mode of coexistence that is always already operative in our relationship with other humans. Indeed, we easily believe that there are common structures of thought that enable us to *know* more about what other humans feel, especially since languages allow us to check whether our guess regarding the other's thoughts was accurate and to refine our sense of empathy. Shouldn't we assert that the dimension of fantasy, in the sense of a pure projection, an arbitrary and vain invention, is more important, if not predominant, in the empathy with animals?

### *Metaphors*

The second classical issue regarding the imaginary of animals relates to metaphors. It is of course correlative to the problem of transposition and of anthropomorphism. Are we not doomed to apply to animals patterns of understanding derived from human experiences? Are the concepts of animals' subjectivity, personality, consciousness, language, and so forth, pure metaphors?

Tim Ingold's analyses in "Hunting and gathering as ways of perceiving the environment" (Ingold 2000, 40–60) give us a deeper insight into such questions. When the Mbuti Pygmies of the Ituri Forest recognize their dependence on the forest that surrounds them by referring to it as "Father" or "Mother" (Ibid., 43), or when the Cree of Northwestern Canada say that animals themselves give them what they need to live, do they metaphorically talk of nature by using patterns coming from the domain of human persons and social relations, so as to make sense of their environment, as claimed by Bird-David (1992, 31) or Gudeman (1986, 43–44)? Ingold rightly argues against this interpretation by underlining its contradictions. (a) First, the hunter-gathers themselves do not characterize their relationships with nature and animals this way (Ingold 2000, 44). (b) Second, they are originally engaged in relationships with other humans, other animals, and the environment: these relations are all bodily, affectively, and spiritually lived. The metaphor theory implies that the relationship with animals and nature would only regard the non-human part of the hunter-gatherers, "since non-human agencies and entities are supposed to have no business in the world of persons" (Ibid., 45): consequently this theory ignores the full-fledged and original engagement that constitutes the hunter-gatherers' relationship with nature and animals. (c) As a result, and this is another argument against it, the metaphor theory, by distinguishing so sharply between nature and culture, actually imposes a Western pattern of interpretation on other cultures while claiming not to do so (Ibid., 44). (d) Finally, the clear-cut distinction between nature and culture upon which the metaphor theory lies is actually highly paradoxical: if, as humans, we live in a cultural world, the very concept of nature (and the distinction between nature and culture) must be cultural. And we can actually observe that it is not shared universally. But the adjective "cultural" itself rests on the distinction between nature and culture. Consequently, asserting that the concept of nature is cultural still needs to refer to an alleged pure nature (that would not be accessible, except through metaphors, to culture), while claiming at the same time that such a concept is a construction (Ibid., 41).

However, Ingold, as well as Jackson (1983)—to whom he refers—and Nadasdy (2007), who claims that Ingold did not go far enough in the overcoming of the metaphor theory, keep thinking within the framework of an exclusive alternative between "metaphorical" and "real." For hunter-gatherers, "animals are not *like* persons, they *are* persons" (Ingold 2000, 51). "Metaphor reveals, not the thisness of a that but rather that *this is that*" (Ibid., 50). "We have tended to treat Northern hunters' conceptions of animals and human-animal relations as "cultural constructions," implying that they are purely symbolic or metaphorical, rather than real" (Nadasdy 2007, 27). "In my view metaphor reveals unity" (Jackson 1983, 6). I do not think this is accurate. In the world of meaning and subjectivity no being "is" actually fully what it is, namely fully self-coincident, mere self-identity. To be sure, a link must exist, otherwise these "metaphors" would not make sense, but what is to be thought is kinship *and* difference between humans and other animals. We also have to think the analogical meaning of the word "person," for instance, when it is applied to humans and to other animals. Speaking of pure identity or regarding the concept of personhood as the basis for a characterization in terms of identity ("animals are persons") are two caricatural traits of Western rationality and positive thought. Even as far as humans are concerned we do not easily know what a person is and if it is really characterized as self-identical. Humans have, with other animals, relations of fascination-repulsion (as shown, for instance, by the word "beast"). The very theory of the opposition between nature and culture shows that it was possible for humans to feel like aliens in this world. To be sure, this opposition is also a construction, somehow a fiction, but it exists as such, has to be explained, and must possess an ontological basis. Metaphors certainly do not create *ex nihilo* fictitious relationships between beings that would have nothing in common; if they did they would never make any sense. But they do create analogies between beings that are not merely identical, and that definitely are not similar in every regard. As a result, there is in effect a process of problematic transposition between humans and other animals, so that the word *metaphor* is consequently not irrelevant, but not in the sense of a mere arbitrary projection: rather in the sense of the openness to and the decentration toward that which is and is not myself. May the "is and is not" be an object of science? Can the study of animals overcome realism without, for all that, losing rigor?

### *Archaic Forms of Meaning*

The question of rigor is also the main challenge put forward by the third main aspect under which the question of the imaginary of animals presents itself to us, namely the issue of the particular nature of coding and expression in animal life. This is, again, directly connected with the two previous issues: if the question of transposition and metaphors can be posed regarding human-animal relationships, it is because (other) animals manifest, through their behaviors and morphologies (see Portmann (1965) on this point), a certain relationship with meaning that, as a result, if not makes sense for us, at least *speaks to us* and provokes us. But this capacity to deal with meanings, to create, to express, and to understand meanings does not take the form under which it is the most manifest in human existence, that is, the form of the language that Merleau-Ponty calls "exact symbolism" (1995, 282): a conventionally established set of well-circumscribed conceptualized meanings, officially linked with precise signs and thus put in common. "Ecological codes do not resemble human linguistic codes or algorithms, but are rather like archetypal imagery or patterns" (Maran 2012, 151). "It is plausible to assume that codes on the ecological level are not strict regulations, but rather ambiguous and fuzzy linkages based on analogies and correspondences" (Maran 2012, 149). These theses are extremely daring and thought-provoking. Thanks to this perspective it becomes possible to better take into account at the same time the kinship and the difference between humans and other animals: analogies and correspondences are not completely foreign to us, they belong to a domain of our psyche that has become marginal in Western thought, regarded as archaic and loose and, when approached through a rational analysis appears as paradoxical, confusing, and even as an epistemological obstacle (Bachelard 1938). Other animals would not then belong to a radically alien world, but we still need to decenter ourselves from our normalized rational thought toward a form of thought that haunts us and remains obscure to us.

Let us explain in more detail this connection between ecological codes and the imaginary. As put forward by Maran, when the expression of a signal is cast in animals' morphology, it may be addressed to other animals (for instance, predators, sexual partners), but these signs are not necessarily perceived by the animal itself: there is no common code shared by the receiver and the sender (Maran 2012, 149). More than that, as noticed by Portmann, in many cases it is impossible to identify a clearly defined receiver; the animal's appearance is then *for X* (Portmann 1965, 234), possibly for various receivers of different species. The close kinship between ecocodes and the human imaginary also becomes patent through the study of the phenomenon of ritualization in animal behaviors[3]: to speak of ritualization amounts to asserting that certain behaviors which *could* be serious are overacted with a characteristic reinforcement of this or that aspect and resumed on a regular basis so much so that they become signs that can be understood in a clear way as such (see, for instance, hunting behaviors ritualized into a courtship display or the broken-wing display used by many birds to lure predators away from the nests[4]). Humans also react in a spontaneous and often unconscious way to sedimented signs, such as squinting as a threat or looking down as a sign of submission. Ritualization is a slow process that takes place at the evolutionary scale, but there still must be some transitional phases during which a serious behavior (achieved in direct connection with the production of a useful mechanical effect) is being converted into a sign without being fully and firmly established as a sign as of yet: it is *becoming* a sign *through ostentation, exaggeration, and separation from effectivity and application to real objects,* namely through perceptible and ambiguous means. As such, these behaviors must then appear as twofold, as what *could* be serious but may be taken as playful (see also Bateson's analyses of the dual nature of playful behaviors: the nip intrinsically refers to the, yet absent, bite, and essentially embodies a relationship to the virtual (1972, 185). In this case the openness of animal behaviors to creativity and, inseparably, ambiguity, is patent. The process of mimicry that takes place at the morphological level and through which, for instance, different eye-shapes, even monstrous and chimeric ones (Langerholc 1991, 192), flourish in an extraordinary variety of species could be analyzed in a similar way. As suggested by Maran, the meaning of these eye-shapes remains mysterious and must be manifold, given the fact that so many species "share" this "code." The latter cannot be absolutely fixed, which can be experienced as well in the fanciful human variations on "natural" warning codes: red and white contrasts, black and yellow stripes, etc. Such indecisive and fluctuating meanings are indeed very similar and akin to this form of thinking which is classically called fantasy and imagination, by contrast with reason: imagination thinks through similarities, analogies, associations, while reason intends to define clear-cut concepts and to identify essential relationships between objects and experiences, instead of simply relying on apparent resemblance. Portmann (1973) also ventures the idea that the sort of expression which can be found in the life and morphology of animals is cognate to human archaic forms of thought or archetypes. Such archetypes surface irresistibly in myths, poetry, dreams, and daydreams. They also contribute to the modeling of our understanding of our environment and of every speech. They were studied and differently defined, for instance, by Jung or Bachelard. Animal codes would then function as human fantasy precisely *because our fantasy would be an inheritance of our animal origins*.

This fascinating theory is, in my opinion, an extremely promising path. Both Maran and Portmann are aware of "walking on thin ice" (Maran 2012, 151), and these theses remain to be more firmly established and rigorously conceptualized. Moreover I want to emphasize that, on the basis of the previous analyses, it is not fully satisfying to characterize this sort of meaning as "fuzzy." Such an adjective expresses an exclusive rationalist perspective that can only define what is beyond it in a negative, privative, and pejorative way, as not completely clear, not rigorous enough. The same goes with the hierarchy that is suggested by the term "archaic." The privileged link between animal studies and the imaginary may appear again as a handicap: may fuzzy meanings become an object of science or of any meticulous investigation? Can animal studies only hope for the discovery of forms of expression and thought that simply correspond to the most elementary, less sophisticated forms of our thought?

I would like to show that the issues posed by the questions of transposition, metaphor, and archaic forms of meaning actually point toward the necessity of rethinking the actual mode of being of animals: questioning our imagination actually must lead to questioning theirs.

## THE IMAGINARY OF ANIMALS TURNED INTO THE ISSUE OF THEIR UBIQUITOUS MODE OF BEING

Let us go back to the transposition problem first. I don't think that the issue at stake is whether we are able to use fantasy to transpose into other animals, but rather whether imagination in this case is the principle of a purely arbitrary and anthropomorphic projection.

Indeed there are, in fact, always elements of projection and invention in empathy, even with other humans. Empathy is fundamentally a form of imagination.[5] By essence the other is not me. Claiming otherwise would be a form of violence. As a consequence, feeling "in the other" what she feels must always be a problematic attitude. Emotional contagion is definitely part of our experience, including in a particularly striking way with animals; I can nonetheless never be fully in the other's place. Empathy also includes elements of assumptions, inventions, and possibly improper projection. There are plenty of examples of misunderstanding, inaccurate fantasies, and radical breaches in human relationships: the actually very frequent accusation of "*bêtise*" (stupidity in French, literally *beastliness*) would suffice to demonstrate it. As a result, the fact that there might be a dimension of imagination in our transposition into animals cannot be a problem in itself: actually using our imagination, and being aware that we do, is the only proper way to understand the other self without reducing it to our identity.

Thus the question is rather: is empathy with animals reduced to mere fantasy, namely a sheer subjective projection? In other words, eventually, as rightly shown by Heidegger, the question of humans' transposition into (other) animals leads to the following one: can my attempts at transposition be welcomed by animals? Is there a correspondence between my use of imagination in empathy and the mode of being of animals? If I can somehow gain access to the other's perspective, while being at the same time myself and outside of myself, if my ubiquitous being increases my capacity for understanding the other, instead of ruining it, then this *other* cannot consist in an in-itself, a finished being; it must, according to Heidegger's phrase, ek-sist (stand out, stand beyond oneself). Ultimately, if Heidegger claims that such a transposition is impossible from humans to animals, it is because he also believes that animals do not ek-sist. As a result, if one demonstrates that other animals do enter into a relation with the virtual and a plurality of perspectives, then empathy must become the most accurate way to understand them.

Similarly, the question of metaphor, as I have tried to show, cannot be whether animals are *or* are not persons. The possibility to think them through metaphors and through analogies with personal relations with other humans demonstrates that they *are* not: they do not possess the full substantial being that is traditionally denoted by the verb *to be*. In other words, I would like to show that they do not possess the mode of being of an inert, self-coinciding, well-circumscribed thing: they consist in an open meaning, liable to evolve and to go through various metamorphoses.

This approach would also enable us to avoid characterizing the expression of meaning in animal life as simply fuzzy or vague, or even archaic. Speaking of fuzziness or vagueness tends to veil a more positive aspect: ambiguity, the capacity for being transposed, "twofoldness," are essential properties of the mode of ek-sistence that characterizes subjects. Moreover, and correlatively, this form of ambiguous or metaphorical thinking has been regarded by several thinkers as more fecund than the rational one, and as what remains at work even within the alleged pure rational discursivity. There cannot be only absolutely determined meanings: the signified is always cut off in a flow of diverse experiences and the boundaries between different objects and concepts always remain unsure. Thus for instance the boundary between plants and animals was always—and still remains—fuzzy. Such a phenomenon leaves room for fluctuations in the usage of words. To be sure, the establishment of and the respect for conventions is an essential condition for an effective communication, but a pure resumption of already existing words and phrases will lead to triviality and insignificance: a dimension of invention always animates languages and builds on the "natural" symbolism of sounds, gestures, matters, etc. (Merleau-Ponty 1945, 203–232). Finally, the first convention could never take place without building on such natural meanings. What is at stake in animal studies is thus the discovery of a common field of invention of meaning, namely a playground in which we may *know* other animals in a new specific sense, not as objects but as ek-sisting subjects. Our imaginary would then become an interface that gives access to the animals' imaginary, the latter being defined not only as a mental faculty, but also and more fundamentally as a creative and ubiquitous mode of being, an existence opened to virtuality and novelty so much so that it cannot be cooped up within this or that objective definition.[6]

In order to establish such a claim, we will need a concept of imagination that does not reduce it to the faculty of creating arbitrary fantasies and, correlatively, a concept of the imaginary that can help us to think the very being of animals. Phenomenology, but also a phenomenological approach to animals' appearance in Portmann's work, will be of great help in this regard, as we will endeavor to show in the next part of this chapter.

## FROM THE HUMAN IMAGINARY OF ANIMALS TO THE ANIMALS' IMAGINARY: A PHENOMENOLOGICAL PATH

I will first explain why phenomenology brought a revolution in the conception of imagination and the imaginary. Then I will study, on the basis of Merleau-Ponty's and Portmann's works, how this renewed conception of imagination can be more specifically connected to animal studies.

### *Phenomenology of Imagination and the Imaginary*

One of the fundamental theses of phenomenology is that appearing is an integral part of the being of things themselves (Levinas 1930, 62). Even when one raises the objection that there could/must be things in themselves, things that are absolutely independent from our representation of them, one actually lapses into a contradiction, since this reference is already a way of thinking such things (Husserl 1950, §9). They are precisely thinkable as inaccessible, as a form of thought that is beyond our faculties: we actually have defined them, and even if one contests this paradoxical definition, this will be again a way of confirming that every object potentially stands within the field of our thought. This representation is a first thin link that makes further investigations possible. Moreover, our world, the world that is accessible to our human mode of thinking and our perceptive faculties, is an open field with indefinite horizons: it already gives us the opportunity to envisage different perspectives, ideal beings and various forms of thoughts. The different imperfect lines that can be found in our surroundings refer to each other[7] and together point toward a horizon of more and more perfect lines, straighter and straighter, thinner and thinner: the ideal geometrical line, which does not belong to the sensible world but helps us to schematize and understand it better, actually still stands on the horizons of this world. Similarly, modifications in our body and empathy with other living beings incite us to think of radically different ways of perceiving the world or even of conceiving it through pure ideas (thus Plato's absolute Ideas are present in our world under the ambiguous form of traces and hints). Everything is thinkable, not in the sense that it would be crystal-clear and perfectly understandable, rather that it always appears even on a distant horizon, and is accessible to a reflection on its meaning, its nature and properties. Its very being must include this manifestation.

To include appearing in things themselves amounts to an introduction of a virus in the classical system. Especially this is true since there are all sorts of appearing, among which are apparitions through images, pictures and fantasies. In the latter, that which appears is given as being present *and* absent, here *and* there. As highlighted by Husserl and Merleau-Ponty, imagination is a form of experience of the object (Husserl 1910, Appendix to the paragraphs 11 and 20; Merleau-Ponty 1961): there may be no connection by nature between the very being of an object and the word that designates it ("dog," "chien," "Hund"), whereas an image (a fantasy or a picture), by contrast, possesses a relationship of resemblance with the object that it represents and essentially manages to make it quasi-present. To imagine is not to abstractly conceive of something, but rather to live an "as if" experience. The Chimera that I imagine or Hamlet on the stage is not present here in the same way as the chair on which the imagining subject is sitting. But they are not entirely absent, nor a mere nothing. Similarly I, who imagine, who experience ubiquity, stand here and there, in my armchair and the Chimera's world. This ubiquitous experience is the essence of imagination. It is tempting to reduce it to pure presence (the imagined object interpreted as being simply an actual neural state, for instance) and/or to pure absence (the imagined object regarded as not being here at all), but that would fail to account for the quasi-presence of the object *through* my state of mind, the postures of my body or my neurons. The very being of the imagining subject, but also of the imagined object, must make possible such a ubiquity. And this regards the being of the imagined object as well, precisely because imagination is an actual experience—of a special kind, but still—of the object: the object manifests itself in its images. And indeed it is impossible to imagine whatever one wants, whenever one wants, and on the basis of any pictorial medium. Some images are more fascinating, more effective than others: they recur over and over again through various daydreams, cultural traditions, myths, and works of art. Some others are empty and insignificant (see, for instance, Bachelard's numerous examples and illuminating analyses in *L'eau et les rêves* [1942] and *L'air et les songes* [1943]). There is a form of transcendence and stubbornness that characterizes the imaginary field. Following Bachelard, I propose to call "the imaginary" this field of images, myths, and fantasies that are not arbitrarily created by a purely subjective faculty, but which haunt our imagination, fascinate it, and incite it to imagine more and more fantasies along the same theme. It is important to make a distinction between the imaginary and the imagination, a subjective faculty that freely creates images: in a classical approach the imaginary is the product of the imagination, hence its transcendence is not acknowledged. In the phenomenological approach that I want to advocate, imagination responds to an immemorial imaginary field that emanates from things themselves as an essential dimension of their very being. It is the object that imagines itself through me. Merleau-Ponty thus asserts that there is an *imaginary texture of reality* (1961, 24): the aspects, properties, characteristics that manifest themselves through our fantasies are actually integral dimensions of the things themselves: dimensions of virtuality, possibility, and ubiquity. Since the imaginary is a field that calls to our imagination, it must consist in a set of open, ambiguous characteristics: this dimension of the imagined objects is deeply manifold, metamorphic, and relational.

Let us now question the specificity of animals in this new conception of imagination. The theory I have just presented is valid for every being; it entails an ontological revolution that I have just outlined and that surpasses the frame of this chapter. But, precisely, animals are not things; they possess a particular relationship to the virtual. Let us examine in what consists the imaginary of animals.

### *The Animals' Imaginary Field*

If the imaginary field always puts forward the power of metamorphosis possessed by every being—things, elements, plants, and animals—what constitutes the specificity of the imagery of animals, is the theme, among others, of human-animal metamorphoses, of proliferating hybrids and mutants. This feature, joined with the themes of speaking animals[8] (in fables, fairy tales, myths, movies, etc.), of animal sociability (see, for instance, Orwell's *Animal Farm*, Kipling's *Jungle Book*, Kafka's *The Mouse*

*Folk*) and of, in a complementary way, on the one hand, veritable friendship and cooperation between humans and other animals and, on the other hand, ferocious threat and aggression, unveils an obsessive motif: that of animals as our most radical *alter ego*. Through them the paradox and the contradiction that tear apart this notion of *alter ego* is brought to a climax. (Other) animals, in the imaginary field, embody extreme proximity and strangeness, the most akin and the most different, the source of a possible extraordinary community as well as of the most terrifying war.[9]

Animals manifest themselves as agents (principles of actions) and radical *alter egos* in our imaginary, which, in the framework of the phenomenological theory of imagination, cannot be interpreted as a figment of our fantasy. Rather: this imagery must find its condition of possibility in the animals' being. As agents and *alter egos* they must possess a specific relationship with the imaginary: like every being their reality is lined with a set of virtualities and open possibilities, but as subjects they present themselves as possessing also an active relationship with the imaginary field; in other words, they possess a faculty of creative imagination, they are able to create new pictures and fantasies which will be acknowledged as coming from them. And indeed ethology and neurosciences study the ability of (certain) animals to, for instance, play (see especially Bateson's (1972, 138-148) analysis of the logical structure of play) and to transpose themselves into another subject's place (Seeman 2011). But it may be in Portmann's work that the most astonishing observations and radical theories in this regard can be found. Indeed, Portmann proposes a systematic study of the relationship between animals, in general, and appearance. The production of images—I will explain why these appearances may rightly be defined as images—is then regarded as an activity that essentially characterizes animals and is at work not only in behaviors, but also at the morphological level. I would like to dwell more on this specific theory first because it gives a striking example of the way biology can evolve into a phenomenology, but also because it shows how Merleau-Ponty, on the basis of the philosophical phenomenological tradition, was able to convert Portmann's hypotheses into an even more radical and daring theory of animals as subjects who imagine and incite us to imagine together with them.

## THE CONCEPT OF AUTHENTIC PHENOMENA IN PORTMANN'S WORK

In *Die Tiergestalt* (1965) Portmann demonstrates that appearances and appearing are not only in the eye of human observers, but are part of the very being of animals. There are criteria that allow us to recognize "organs to be seen," developed by animals themselves (and that should be distinguished from animals' appearances, which simply happen to seem beautiful or meaningful *to us*): they are especially schematic, readable (Portmann shows, for instance, that hidden organs are disposed in a chaotic way while animals' external appearance is symmetrical and makes possible an easy identification of the species), they present vivid colors, patches of colors outlined with black, mimicking patterns, and so forth, they are obstinately repeated in different individuals of the same species or by the same individual through time, they are able to deceive the observer, or to warn it, or to make the animal recognizable or appealing for its sexual mates, etc. Moreover, these patterns are the product of systematically oriented morphogenetic processes that imply and combine diverse elements of the body so much so that they compose one and the same global appearance (see the famous Oudemans' principle and Portmann 1958). Thus animals themselves, as subjects, actively implement a masterful art of appearing. Portmann called "authentic phenomena [*eigentliche Erscheinungen*]" those appearances produced *by animals* which cannot be reduced to idle anthropomorphic fancies: as objective realities, they can be studied scientifically, but, at the same time, they are intrinsically the expression of an oriented process, an intentional and subjective process—Portmann follows von Uexküll on this point. Portmann thus turned expressive animal subjectivity into an object of science. But he went even further, first by showing that such authentic phenomena can be found also in lower forms of animal life, in animals that do not possess eyes, or who live in the dark and cannot be seen by any mate or prey or predator—even though they become more and more thematized and sophisticated in more highly developed animals (1965, chapters 1 and 2). These are "unaddressed phenomena [*unadressierte Erscheinungen*]" (Ibid., 234), "sent "into the blue" [*"ins Blaue" gesendet*]" (Ibid.): there are innumerable possible receivers and the meaning of the "signal" becomes virtually multiple.

Moreover, Portmann insists on the idea that it is, in general, impossible to connect such authentic phenomena exclusively to a determined function. To be sure, in certain cases, the appearance produced by an organism obviously triggers marked reactions in other specific animals: the eyes drawn on the wings of butterflies, for instance, when displayed suddenly, very efficiently set off a momentary state of stupor in possible predators. In such cases, the phenomenon seems to be designed for producing a defined effect: it is destined for, and perfectly adapted to the eye of a determined receiver. But, as Portmann points out, in many other cases, and even in these cases, the function, as well as the receiver, cannot be so easily and clearly identified. Predators get accustomed to the trick (their stomachs are full of mimetic animals) (Caillois 1938, 105; Langerholc 1991, 190, 199), some appearances are ambiguous (Kleisner and Markoš 2005, 5–7), many of them are useless or even damageable (Portmann 1965, end of chapter 9, the example of the gonads, Grene 1965). What is more, neither the vague concept of utility, nor the alleged survival instinct, nor the model of the selection of the fittest can account for the fantastic and fanciful diversity of appearances that are produced by animals. A code, a system of signals would be more effective if it were more limited and more regular. Portmann thus presents various hypotheses regarding the meaning and the function of animals' art of appearing: appearing might be "a basic property of life [*eine basale Lebenseigenschaft*]" (1965, 233): living beings would appear and create new appearances for the sake of appearing and playing with appearances. In other words life would serve the art of appearing and not the contrary (Portmann 1996, 160). Appearing might then be autotelic and self-satisfactory, like play, which can be actually experienced in everyday human existence through narcissism, fashion, heraldic practices, art, etc. Animals' creation of appearances might also be regarded, Portmann suggests, as the way to proclaim, so to say "to the world," one's proper, individual, or specific, value (1965, 232). Portmann does not make clear whether the intentionality that runs such appearing processes aspires to find recognition, to express the interiority of the animal (see the concept of *Selbstdarstellung* in Portmann 1965), or to arouse the pleasure and the joy of contemplation that it actually often provokes, at least in human observers (which Portmann also wants to take into account and to consider as an integral part of these phenomena, see 1965, 239–240).[10] Thus Portmann also puts forward the kinship between this animal art of appearance and human art or human fantasy: "At times, the sight of these organic forms makes us feel as if we are faced with the uncanny materialization of our dream life, the products of our fantasy" (1965, 240; Eng. tr. 106)[11]. Portmann highlights that human symbolism abundantly resorts to figures and forms borrowed from animal appearances (Ibid., Conclusion), but he also acknowledges that the idea of an intrinsic connection between the latter and human imagination is still a sheer "inkling [*Ahnung*]" that deserves attention but "cannot be more firmly founded for the moment" (Ibid., 240). In fact, Portmann outlines different speculative assumptions, but also emphasizes that the role of science is to avoid letting any assumption or metaphysical option limit the access to the richness of phenomena (Ibid., 65).

Furthermore, another aspect of Portmann's theories hinders the defense of the claim that the production of appearances in animals follows the same motivations as human art or human fantasy: in his anthropological texts Portmann maintains a clear distinction between humans and animals. When biology moves from animals to humans, Portmann remarks in "Um eine Basale Anthropologie," it must be struck by the openness of the structures of existence and the "freedom" (1973, 121, 279) to endlessly invent new forms of existence. In animal life "on the contrary," even for more highly developed species, behaviors as well as appearances are hereditarily determined (Ibid., 279). How could we not become suspicious again, then, regarding the legitimacy of considering our daydreams, our playful interpretations, and fantasies as integral parts of the meaning of animals' appearances? On this point precisely, I think, phenomenology can help us to take a step further.

## MERLEAU-PONTY'S READING OF PORTMANN

Merleau-Ponty devoted several passages of the *Nature* lectures to Portmann and drew considerable inspiration from his work, but he was much more assertive than Portmann was regarding the connection between animals and fantasy. Merleau-Ponty uses the phenomenological approach precisely as the lever that was missing in Portmann's work, and makes possible the establishment of an essential continuity between imagination and animals' appearances. Merleau-Ponty's text is entirely structured by the pattern of reciprocity: there is an *Ineinander* between humans and other animals—in other words, we live through them and they live through us—and the animal can be seen "as variant of humanity and humanity as variant of animality" (2003, 307), human language is a "resumption [reprise]" of the forms of expression developed in nature and living beings (Ibid., 219), and "human desire emerges from animal desire" (Ibid., 288). We should "extend to animals what Descartes says about the human body, defining it as this sort of body that cannot be closed on itself in the manner a fragment of space may be, because life teaches us not only the union of our soul and our body, but also the lateral union of animality and humanity" (Ibid., 271). The implicit argument that justifies such assertions is, I think, the typical phenomenological approach to appearance and fantasy. Our representation of animals as expressing oneiric meanings even through their morphology cannot be a mere projection: it must have been instituted by the being of animals itself. Thus Merleau-Ponty insists on the idea that animals themselves and their peculiar nature *forced* a traditionally objectifying and analytical science to take into account subjectivity, global *Gestalten* and "fuzzy" open meanings (he mainly refers to von Uexküll, Lorenz and Portmann). "Life gives itself only to the one who seeks to see, not to grip the solid core" (Merleau-Ponty 1995, 303). Animals have genuinely decentered us: through our representation of them we have been carried away into disconcerting experiences and stubborn fantasies. And Merleau-Ponty emphasizes the fact that what is instituted within us by animals takes the form of imaginations, transpositions, metaphors, myths, and questions. Merleau-Ponty refers in particular to Eskimo zoomorphic masks (Ibid., 270, 277) that manifest the human desire to transform into animals: this desire itself should be taken as the proof that we find, in the very being of animals, a call, the prompting to seek in it aspects of ourselves that we essentially need to find so as to better understand and accomplish ourselves. This desire would not exist under this obsessive form if our being were devoid of kinship with animals and if animals' being were not any longer able to support and feed this impression of kinship. Humans keep returning to animality for questioning it about its own origins and its own nature. Acknowledging that these questions exist, keep haunting us, and remain open, actually leads us to deduce that animals are first and foremost a source of questions and speculation, namely that they cannot be reduced to a closed set of molecules and functions. "It is not a positive being but an interrogative being which defines life" (Merleau-Ponty 2003, 156). Therefore animals must be defined as the institution of an open imagination. Through myths we never do get positive and definitive answers: existence cannot be reduced to equations and formulas. The *quest* instituted within us by animals cannot be, at its source, launched by a set of positive and closed meanings: otherwise such a questioning would not be aroused *by animals*, by the consideration of animals, and would not be a *stubborn*, despite ourselves, *ever-recurring* questioning *about* animals, but rather an arbitrary and contingent delusion, devoid of privileged connection with animality. If we are genuinely decentered by the consideration of animals, therefore animals too must be decentered by our representation of them: human myths and fantasies are the continuation of the expressive processes that were started by animals.

Portmann's thesis that animals send appearances to the blue can then be taken fully seriously and better understood: "Animals (Portmann): the body as organ of the *for-other*" (Merleau-Ponty 1995, 272, 281, 288): animals, as forms of expression that can be resumed by humans' questions, desires, and imaginary fields, must be essentially opened to and plugged onto the intersubjective and interspecific field where ambiguous meanings live their lives and are shared, resumed, and modified by subjects of all species. This field is a playground rather than the rational community based on universal concepts. Correlatively it can be asserted that animals essentially produce ambiguous meanings that Merleau-Ponty defines according to the model of a melody: they are not simply vague or fuzzy meanings, they are rather like musical themes (Ibid., 239) that lend themselves to different variations and that cannot be properly understood if they are not creatively resumed. "The species is not so much behind animals, as ahead of them as an *Ineinander* (Portmann)" (Ibid., 288, my translation[12]). This claim can, I think, be expanded: it is not only the species that is ahead of animals (namely that remains to be defined through the behaviors and forms that will be produced by individuals, that will be exchanged, mimicked, ritualized, etc.), it is an interspecific field of meanings, since different species also exchange forms and meanings and since the boundaries of species are not rigidly established. Consequently animals *are* not, they have their being ahead of them. As a result, Merleau-Ponty puts forward the claim that animals' being is comparable to a melody that sings itself or to a "pure wake that is related to no boat" (2003, 176): it cannot be defined as a perfectly circumscribed plan, a positive essence. In this sense such a being is a form of imagination: animals do not consist in any positive reality: they must be here and there and haunt many different forms of embodiment and expression (see also for a more concrete description of this interspecific field of meanings in Kleisner and Markoš (2005) on the concept of *seme*). Thanks to a dynamical definition of humans and other animals as beings that are ahead of themselves, as ubiquitous, essentially open and relational beings, it becomes possible to escape a positive mode of thinking that always fears to import into the concept of animals elements that would be strictly proper to humans and *vice versa*. Differences between humans and other animals can be taken into account, but do not prevent us any more from thinking the kinship between them and, more than that, their partaking in a field of meaning ruled by invention and variations on open themes.

The more we try to fix the nature of animals and to hold it firmly between the pliers of positive concepts, the less imagination will be invited to take part in the study of animals, and the less animals' imaginary dimension and active imagination will be recognized and known. This does not mean that science should turn into a series of purely playful fantasies. More fruitfully, the existence and the vivacity of this imaginary of animals may be acknowledged as a meaningful phenomenon in itself that can be analyzed and which teaches several fundamental theses regarding human and animal nature. Moreover, as pointed out by Maran (2012), it also becomes possible to study the form and the content of the themes, archetypes, or images that circulate from species to species: they cannot be understood along the model of positive or exact language, nor according to a purely mechanistic approach (because of their ambiguity and ubiquity): the model of human imaginary, of archetypes, symbols, and dreams could here become of great help and part of a new paradigm.

## NOTES

1. This text was written as part of the grant-funded project GACR "Life and Environment. Phenomenological Relations between Subjectivity and Natural World" (GAP15-10832S). I also want to thank Louise Molly Westling, who is Professor Emerita of English and Environmental Study at the University of Oregon, for her great help in correcting my English as well as for all the illuminating discussions without which this paper would not exist. I am of course fully responsible for all the mistakes that remain in the text.
2. "Despite Thomas Nagel, who is probably a good man, despite Thomas Aquinas and René Descartes, with whom I have more difficulty in sympathizing, there is no limit to the extent to which we can think ourselves into the being of another. There are no bounds to the sympathetic imagination" (Coetzee 1999, 133).
3. See in particular Haldane 1953, Huxley 1966, and McFarland 1999, 377–400.
4. See, for instance, Caro and Girling 2005, 343–347.

5. More precisely, Husserl defines it as a form of presentification [*Vergegenwärtigung*], an act through which I make present what is not; see Husserl 1954, 211, 280.

6. See also on this point Vicki Hearne 1987.

7. This synthesis cannot be the pure product of our mind; otherwise it would not give us access to any obdurate and transcendent reality, and everything should be crystal clear to us. Even in a Kantian point of view, there must be a sensible matter to which our syntheses apply and that must present some motivations that justify this or that conceptual shaping so much so that a world instead of a mere arbitrary and flimsy fantasy emerges.

8. Let us add the theme of the secret and immemorial wisdom possessed by (some) animals and that they can conceal from humans or pass along to them (for instance, Nietzsche's *Thus Spoke Zarathustra* or Alfred Adler 1998, §48).

9. The theme of devoration conveys similar traits: animals are not simply forces of nature that may destroy me, but rather what tends to appropriate and assimilate what it encounters. Animals live also from the assimilation of the other, first through food assimilation and, more specifically, by contrast with plants, the assimilation of organic matter (the process of destruction and ingestion of the akin is particularly manifest in this case), but also through the institution of an *Umwelt* that exists only for it, as shown by von Uexküll. Animals establish a precarious equilibrium between the opening to the other (to a world, to other individuals and species), adaptation to it and the modification or even assimilation of it that goes together with the experience of its vulnerability. Of course, a more systematic study of the main themes that constitute the imaginary of animal remains to be completed.

10. Portmann also considers the idea that such a *Selbstdarstellung* might be an archaic form of reflection, a sort of communication to oneself, similar to the one that we can experience when increased heart rate, sensation of smothering, warmth in our face inform us about our emotional state (Portmann 1973, 286)

11. "Zuweilen ist uns im Anblick dieser Gestalten zumute, als begegneten wir den Ausgeburten unseres träumenden Lebens, den Erzeugnissen unserer Phantasie."

12. "L'apparence des animaux (robe, ornements) comme organe de communication, l'espèce n'est pas tellement derrière eux que devant eux comme *Ineinander* (Portmann)."

# FIVE

# Speaking with Animals

## *Philosophical Interspecies Investigations*

*Eva Meijer*

Language is usually seen as a human endeavor, in philosophy, societies, and politics. Other animals are regarded as mute because they do not use human language, even though they clearly communicate amongst themselves and with humans. In the Western philosophical tradition humans are viewed as radically different from other animals, and language is seen as one of the main ways in which this difference is expressed. It is used to demarcate the border of who is human, and therefore the concepts "language" and "animal" are repeatedly defined in relation to one another. Humans are also separated from other animals in language, by using the word "animal" to categorize all nonhuman animals. Jacques Derrida formulates this relation between exclusion and speech as follows: "Men would be first and foremost those living creatures who have given themselves the word that enables them to speak of the animal with a single voice and to designate it as the single being that remains without a response, without a word with which to respond" (2008, 32).

The Anthropocene, understood as the epoch that began when human activity started to significantly affect ecosystems globally, is interconnected with human exceptionalism. Viewing humans as fundamentally different, and hierarchically above other animals, leads humans to act in ways that do not take into account the well-being of other animals, and to see their interests as less important than human interests. The term Anthropocene also implies that as a species, humans are different from other animal species—for better or for worse. However, while human impact on the planet increases, human exceptionalism is increasingly challenged. Animal rights theorists argue that we should view other animals as subjects with their own perspective on life. Recent work in political philosophy proposes to view nonhuman animals as political actors, which shifts questions about animals from how we should treat them to how we can gain insight into how they want to live their lives, what types of relationships they desire with one another and with humans, and so on (Donaldson and Kymlicka 2011). Poststructuralist and posthumanist thought challenges the binary opposition between humans and other animals (Calarco 2008; Derrida 2008; Wolfe 2003). Recent empirical studies in biology and ethology that focus on cognition (Allen and Bekoff, 1999), cultures (Smuts 2001), and languages (Gentner et al. 2006; Slobodchikoff et al. 2009) of nonhuman animals argue for seeing the difference between human and nonhuman animals as a difference of degree, not kind.

If we view other animals as subjects with their own perspectives on life, something the abovementioned authors convincingly argue for, it is not enough to think *about* them; we need to start thinking *with* them, in order not to repeat anthropocentrism. This means rethinking both our relations with nonhuman animals and the concepts attached to those relations. In thinking about thinking *with* other animals, language plays an important role. Learning about other animals' languages can help us understand them better, and build new relations with them; challenging an anthropocentric view of language can help us see animals of other species, and their languages, differently. Conceptualizing animal languages can also play an important role in addressing certain practical problems between humans and other animals in the Anthropocene, such as border conflicts between groups of wild animals and human groups, or the position of domesticated animals in human societies.

In this chapter I critically examine the existing anthropocentric view of language, and use concepts developed in the phenomenological tradition, and in the later work of Ludwig Wittgenstein, to think about animal languages and interspecies worlds. I first briefly discuss the dominant view of language and animals in the philosophical tradition, as well as Derrida's critique of this view. I then focus on conceptualizing existing human-animal languages, building on Wittgenstein's later view of language, in particular his concept language-games. Finally, I investigate the relation between language and world, and explore possibilities for building new languages and worlds with other animals. The aim of these investigations is not to give a new definition of language, but to explore alternatives to an anthropocentric view of language, and to investigate the connection between language and world in interspecies relations.

## THE ANIMAL AND LANGUAGE

In Book I of *Politics,* Aristotle defines man as a political animal and the only animal that is endowed with speech. By doing so, he draws a line between humans and other animals. This line functions as a border around the political; only humans can be political animals for they possess reason. It also defines what counts as speech, or reason: nonhuman animals have voices but do not express themselves in human language, and cannot decide between just and unjust. What is furthermore important to note is that Aristotle uses human words to argue that using human words is what determines the border between who can speak and who cannot, and between who can be a political actor and who cannot.

Later philosophers also stressed the interconnection between *logos* and language. René Descartes (1644, 1646) argued that the fact that animals do not use language is proof they do not have thoughts; even animals who do speak, and here he gives the example of magpies, do not express thoughts, they simply express a passion. Because animals can express passions in words, they should be able to express thoughts in a similar way if they had any, but they do not, which shows that they do not think. Their passions are mechanical and thus animals can be compared to machines; Descartes calls them bêtes-machines. Because these bêtes-machines do not possess reason they do not feel; they function like clocks and their responses to pain are purely mechanical. Immanuel Kant did not deny that animals feel pain, although he also argued that animals—as well as, to some extent, certain groups of humans, such as children and women—do not possess reason and therefore are not moral subjects. Because other animals are not rational beings, they are deprived of liberty, autonomy, and dignity, and they cannot be the subject of rights and duties. An animal can never be an "end in itself," but only a means (Kant 1997, 428). Kant does not explicitly discuss the language of nonhuman animals, but remarks that to be a person one needs to be able to have the representation "I," something that is often shown in language (Kant 2007, 127). Martin Heidegger (1929) argues that the animal is "poor-in-world"; animals have access to the world, they experience it, but they do not have access to the world *as such.* They are captivated in their environment; they cannot break out of the ring that forms their environment because they are absorbed by it. In contrast, (human) *Dasein* can relate to the world *as* world, to the Being of things, and to itself as *Dasein*. Because the animal lacks the *as such,* she also lacks logos and connected to that language, since logos is founded on (or in the possibility of) the *as such.* The difference between nonhuman and human animals is not one of degree but one of kind; according to Heidegger they are separated by an "abyss" (1929, 248).

These different interpretations of humanism see language, understood as human language, as what creates or expresses a pure break between human and nonhuman animals. This presumes a specific view of language, as the expression of objective, determinate meanings (Glendinning 1998, 77), which humans alone can grasp with reason. According to Derrida (1992), the idea that there is an objective meaning to words is the main problem of philosophy, because it always involves exclusion and isolation of possibilities in the phenomenon of language, and consequently in the outer world. In his own view of language, he emphasizes the shifting of meanings. His concept "play" refers to the structure of systems of signs, more specifically: "the spacing between the pieces of an apparatus [which] allows for movement and articulation" (Derrida, 1992, 64; see also Glendinning 1998 chapter 5 for a discussion). Although this mechanism can give rise to misunderstandings, it is needed for a language to function normally. Instead of conceiving of meaning as univocal and objective, in studying language we need to take into account its structural ambiguity.

In his later writings about animals Derrida (2008, 2009, 2011) follows "the animal" through the Western philosophical tradition and argues that the animal functions as an Other or negative to the construction of the human subject. In this construction, language plays a prominent role. Nonhuman animals are defined by their lack of language, and language is defined as solely human, which deprives other animals of a voice in human societies. Also, all other animals are categorized into one group by the word "animal," which allows humans to disregard their lives and interests more easily (2008, 31). Derrida argues that a clear separation between humans and animals is problematic because it obscures the many differences between individuals and groups, of human and nonhuman animals. "I am trying to explain how drawing an oppositional limit itself blurs the differences, the difference and the differences, not only between man and animal, but among animal societies—there are an infinite number of animal societies, and, within the animal societies and within human society itself, so many differences" (1995, 32). Instead of defining one line between Man on the one side and Animal on the other side, Derrida argues that there is a multiple and heterogeneous border (2008, 31). To account for this, Derrida proposes the word "animot." In speaking form it refers to the plural, the multiplicity of animals, which is necessary because there is not one "animal." It shows us that it is a word for animals, not a referent to an existing group of animals; the "mot" in animot refers to the act of naming and the stakes involved in drawing a limit between human and animal, by the human (2008, 31).

Derrida does not develop a positive theory of language which includes other animals, because extending the humanist framework on which our concept of language is founded to incorporate other animals would leave the ontological structure, based on the (metaphysical) distinction between The Animal and The Human, intact. Inherent in this would be the risk of repeating violent consequences (Wolfe 2003; Calarco

2005; Oliver 2007). Instead, we need to challenge this structure of thinking, and *think* new ways of thinking: "Hence the strategy in question would consist in pluralizing and varying the 'as such,' and, instead of simply giving speech back to the animal, or giving to the animal what the human deprives it of, as it were, in marking that the human is, in a way, similarly 'deprived,' by means of a privation that is not a privation, and that there is no pure and simple 'as such'" (2008, 160). While this analysis of the *as such* is relevant in rethinking the capacities of human and nonhuman animals, it seems problematic to state that we should not give speech back to other animals, because this keeps intact the hierarchy in which the human decides who speaks. It is also a fiction to think that we can *think* the absence of language. The question should therefore not be whether the animal can speak, or whether we should allow her to speak, but what she tells us, how we can find this out, and how we can respond.

It is important to recognize that concepts are already influenced by other animals. Humans learn the meaning of words and concepts in interspecies contexts, where they are influenced by and formed in interaction with animals of other species. In his discussion of anthropomorphism, Raimond Gaita (2002) argues for taking the public character of language into account in interspecies communities. He gives the example of intentions (2002, 60) and argues that it is wrong to think this is a human concept that we apply to other animals. Rather, the meaning of "intention" is formed by interaction with, and influenced by the behavior of, different kinds of animals; humans learn what "intention" means by watching the intentions of animals and humans, by reading about the intentions of others in books, and by understanding one's own intentions as intentions. In this view, saying that animals have intentions is not anthropomorphic. This is not due to its truth-value in a judgment of the mental states of other animals, but related to the fact that animal intentions are part of what gives "intention" its meaning. In this context, Kersty Hobson emphasizes the material-political dimensions of living with other animals, and argues that animals have co-created "the histories, moralities, political subjectivities and places we take as natural and/or devised through human ingenuity alone" (2007, 257). Acknowledging the presence of other animals in societies and concepts is, however, not sufficient for rethinking language. For that we need to engage with actual nonhuman animals.

## ANIMAL LANGUAGES AS LANGUAGE-GAMES

The Border Collie Chaser learned and retained the proper-noun names of 1022 objects: 800 cloth animals, 116 balls, 26 Frisbees, and a medley of plastic items, and learned to categorize these objects over a three-year period of intensive training (Pilly and Reid 2011). Later research showed that Chaser can learn by deduction and understands grammar (Pilley 2013). It is clear that Chaser has a good memory; psychologist John Pilley had to write the names on the toys with a marker because he couldn't remember them. It is also clear that she is a fast learner. Critics, however, argue that it is unclear what the relevance of these studies is for researching the foundations of language, or learning a language, since Chaser only learned to perform tasks by repetition. This is as much, if not more, a question about what language is, as it is a question of what Chaser does and does not do when she relates to words.

In *Philosophical Investigations*[1] (1958), Wittgenstein argues that we cannot give a single definition of language: there are many different ways in which we use language that are related, but these do not share one characteristic, so therefore there is no one way to describe them. Instead of trying to find a definition, we should describe and investigate these different uses of language, which he calls language-games. "Instead of producing something common to all that we call language, I am saying that these phenomena have no one thing in common which makes us use the same word, but they are related to one another in many different ways. And it is because of this relationship, or these relationships, that we call them all 'language'" (*PI*, §65). Wittgenstein does not give a clear definition of language-games. He uses the concept to refer to the whole of our natural language, as comprised of a collection of language-games, but also to simple examples of language use—such as Chaser's learning of words (*PI*, §7). Wittgenstein also uses this concept for the most primitive forms of language (*PI*, §7), which are not only available to humans (*PI*, §25; Glendinning 1998). "(I) will sometimes speak of a primitive language as a language-game. And the processes of naming the stones and of repeating words after someone might also be called language-games. Think of much of the use of words in games like ring-a-ring-a-roses. I shall also call the whole, consisting of language and the actions into which it is woven, the "language-game"" (*PI*, §7). Language-games are open-ended, there is always a possibility for the realization of new language-games, and there are many language-games we do not even recognize as such: "We remain unconscious of the prodigious diversity of all the every-day language-games because the clothing of our language makes everything alike. Something new (spontaneous, 'specific') is always a language-game" (*PI*, II xi: 224). Furthermore, Wittgenstein emphasizes the importance of gestures and other non-linguistic expressions in language-games (*PI*, §7) and he compares them to tools (*PI*, §11). To grasp what language is, we should study language-games, by studying the practices in which they take place.

Humans and other animals have many different relationships and encounters and animals express themselves in many different ways. Thinking of language as a collection of language-games is an appropriate starting point for thinking about interspecies languages, because it does not discriminate between different types of linguistic and non-linguistic acts, and because it emphasizes the relation between meaning and use (see also Meijer 2013). Wittgenstein shows that language is essentially a public practice. Recognizing that meaning is created and located in the relation between language and practices, not mind and language, avoids the pitfalls of humanism I discussed in the former section. Instead of asking ourselves whether other animals possess a true language, informed by logos and located in the mind, we can investigate human-animal language-games by studying the practices in which they take place.

Between humans and dogs, we find many different interactions. These can include the use of human words, gestures, eye contact, smell, and other forms of communication (Hare and Woods 2013). Some of these forms of communication are more complex than others, similar to interaction between humans. Chaser's and Pilley's use of language seems to be very similar to the Augustinian picture of language that Wittgenstein discusses in the beginning of the *PI*. Chaser learns to stick words on objects, and every word refers to one object. But this is not the only language-game that Chaser takes part in. The language use of Chaser and Pilley can be seen as a cluster of language-games. Learning to put words on objects is a language-game, as is learning to categorize objects, the interaction with Pilley in the learning process, and so on.

Animal trainer and philosopher Vicki Hearne (1994) describes how she teaches the pointer Salty to fetch a dumbbell. Hearne sees this as teaching her the language-game "to fetch." Salty learns what it means to fetch something and according to Hearne this enables her to express herself more fully, for example by making jokes. Salty jokes by fetching the garbage bin or a car tire or she fetches the dumbbell but brings it to someone else. When she does this, her bodily movements (such as wagging her tail) and facial expression are playful and joyful. Teaching a dog the meaning of a word clarifies the interaction between dog and human, both on the side of the dog and on the side of the human, and lays the basis for further interaction. The word "fetch" gains its full meaning in being taught to the dog, and afterwards the dog and human can use it when they interact. Between dogs and humans we find language-games that involve the use of human words (yes, no, eat, ball, good, here), but using human words is not necessary. Humans and dogs may use their voices, body language, eye contact,[2] gestures, or movements. Communication may vary over time, as the relationship grows and understanding evolves. A dog-human relationship is not the same as a human-human relationship, yet it has its own authenticity. Hearne discusses training and working situations, in which "genuine animal partners of humans (e.g., dogs and horses) become "kinesthetically legible" and vice versa" (Acampora 2004). This partnership is made possible by both human and dog, and brings into being not only a practical understanding but also a moral framework. Within the shared language-games, dog and human can be held accountable (Hearne 1994).

### *Speaking Animals*

Animals of different species express themselves differently. To communicate they use scents, colors, movements, words, songs, sounds, patterns on their skin, and so on. Humans are only beginning to understand the meaning of many of these signs and expressions. Recent research has, for example shown, that dolphins (King and Janek 2013) and parrots (Berg et al. 2011) call each other by their names. African Elephants use different alarm calls to distinguish between threats from bees and humans (Soltis et al. 2014). Ravens use referential gestures (Pika and Bugnyar 2011). Fork tailed drongos mimic the alarm calls of other species to scare animals away and then steal their dinner (Flower et al. 2014). In the songs of many species of birds we find grammatical structures, including recursion (Gentner et al. 2006). The skin patterns of squid can be seen as a language which includes grammar (Moynihan 1991). Prairie dogs describe intruders in detail, including the color of their T-shirts (Slobodchikoff et al. 2009).

Other animals also share a variety of language-games with humans. I already discussed humans who teach dogs the meaning of human words. Irene Pepperberg (1995) investigated the linguistic capabilities of the African Grey Parrot Alex. Until then, human-parrot communication had been limited to the use of simple words, because humans and parrots learn languages differently, and use language in different ways. For parrots, learning a language is strongly interconnected with acting, and Pepperberg made use of this in the interaction with Alex. For example, she gave him control over his rewards. If he recognized objects he could either have them, or choose something to eat, or go out for a walk. Through developing different language-games in which words and actions were intertwined, he learned—among other things—not only to recognize, describe, and categorize many objects, but also to use concepts such as "same" or "different," and commands to control the behavior of others ("come here," "go away"). The new language that parrot and human created offered Pepperberg insight into Alex's mind, for example, when one day he asked Pepperberg what color he was.

In other language experiments dolphins were taught to speak in human words, and nonhuman primates were taught to speak, sign, and use keyboards and symbols. Famous examples of animals with large vocabularies include the chimpanzee Washoe and the gorilla Koko, who both used sign language, and the bonobo Kanzi, who uses signs and lexigrams to communicate with humans. Sometimes nonhuman animals start to imitate human speech at their own initiative. The beluga whale Noc, who worked for the US army, copied human chatter. Elephants can use their trunk to manipulate the sounds they make, which allows them to pronounce some human words. The Asian elephant Batyr started speaking in this manner in a zoo in Kazakhstan, where he was held captive his whole life, without the company of other members of his species.

Experiments in which nonhuman animals are taught to use human words can provide us with information about their learning behaviors, especially regarding mimicry, and about their attunement to humans. They can in some instances give us insight into their minds, for example, when Koko signs about her memories. But they do not tell us everything or even much about their linguistic skills; at best, human language can function as a tool to develop a new type of language, as with Alex and Pepperberg. Similarly, while speaking elephants and whales show us that they are creative and social beings, their imitations of humans do not tell us much about the languages of whales and elephants. Both species have rich and complex languages, which we are only just beginning to learn. The concept language-games allows us to see that experiments in which other animals are taught to speak in human language are part of a specific family of language-games, in which only human language is seen as language, and in which humans set the terms for interacting in a scientific context. Noc and Batyr both create new language-games, in order to connect with the humans around them. Human words play a role in some, but not all, language-games. In his discussion of aesthetic judgments, which he considers to be complex and refined, Wittgenstein emphasizes the importance of gestures (1978). A slight move of the head, a nod, might be a better expression of a judgment than a word ("beautiful"). In the languages of other animals, gestures, movements, and other nonverbal expressions play an important role. We can only study these languages properly if we let go of the idea of human language as the only true language.

## BUILDING INTERSPECIES WORLDS

Ethologist Konrad Lorenz shared his life with many animals, and describes communication with them in detail (1949). For example, he observed parrots and corvines who learned to use the right word at the right occasion (i.e., who said "good morning" only in the morning). He connected this to a strong sensitivity to human behavior, although in some cases, when experiences make a strong impression, the birds can learn to use words themselves, after hearing them only one or two times (1949, 84). For some birds, such as mallards, sound is decisive for choosing whom to follow. To get the mallards he raised to follow him, Lorenz needed to copy their mother's call-note (1949, 40). He managed to invent a call-note that both parties could understand, thereby creating a new interspecies language-game. Another example of a shared call-note concerned Roah, a raven who had been with Lorenz since he was young and remained a companion in later years. Roah often accompanied Lorenz on walks and when he went skiing. As he got older, the bird became shy of strangers and of places where he once had had unpleasant experiences. He did not like it when Lorenz lingered in these spots, and when he did, Roah flew over his head, as jackdaws, for example, do to make their children leave the ground and fly with them, and called him. The call-note he used for this was the call-note Lorenz used for Roah: his own name, with human intonation. Roah only used this call for Lorenz, when addressing one of his own species, he used the call-note they normally used in this group. Lorenz argues that he cannot unconsciously have trained the bird; for this he would have had to come up to Roah at the moment when he was calling Lorenz's name and at the same time wanting his company, which would have to have happened at multiple occasions to create a connection. So, Lorenz argues, Roah must have possessed the insight that "Roah" was Lorenz' call-note.

Len Howard (1952, 1956) shows that to create understanding it is not necessary to raise wild animals, as Lorenz did. Howard believed that the best way to study birds was on a basis of trust and in a natural setting, instead of in laboratories where the birds were held captive, as was common at the time. Captivity influences the birds' behavior; it makes them nervous. To avoid this, Howard opened her cottage in Sussex—literally; she kept the windows open—to the great tits, robins, sparrows, blackbirds, thrushes, finches, and other birds that lived in the area. Birds are curious and they soon came closer, flying in and out of the house as they pleased. They quickly learned not to be afraid of Howard. She made nesting places for them in the house, and they also made their own nests inside. Howard lived with the birds for many years and got to know them intimately as individuals. She carefully describes their behavior and argues that they act on intelligence, not just instinct. Trained as a musician, she also studied and wrote down their songs. The communication between Howard and the individual birds included gestures, eye contact, tone of voice, and sometimes also human words, as becomes clear in the following description of great tits who want to peck at Howard's butter dish, which they know is usually forbidden:

> (T)hey perch a little way off and look first at the butter, then at my face, hesitating although longing to help themselves, for they have a passion for butter. If I say "come on" coaxingly, they confidently step up and eat it. If I say "no" just a little sternly they remain where they are, but continue to look pleadingly at me and then the butter. A shade crosser "no" sends them hopping farther off; an angry "no" makes them fly to the open window, but if I quickly call out, "Here, come on," in a very coaxing tone, they at once return and if I keep quiet, inch by inch they hop along the table towards the butter, still eyeing me for further signs of objection. . . . They interpret correctly any sign of objection in voice

> or movement, but without the encouraging tone of voice they will not touch butter while I am looking because I have once or twice forbidden them by an angry "no." (1952:19)

Howard and the birds that lived in and near Bird Cottage show how new forms of language can be constructed between communicative beings of different species. Richard Iveson (2010, 2012) draws on Heidegger's view of language to show how human and nonhuman animals create meaning. For Heidegger, *Dasein* is constituted within infinitely entangled structures of meaning, which *Dasein* can only notice when it is thrown out of it in a state of Angst or boredom. In these states, *Dasein* moves from concealment to an authentic existential experience. According to Heidegger, the animal cannot experience this because she lacks language. Iveson argues against this, because nonhuman animals are thrown into a world that exists of meaning giving structures as well. They are constituted by and influence these structures, and language is primordial with this. "Language" here is not one true language, but rather the constructed, artificial view of language which Derrida puts forward in *The Beast and the Sovereign II*. Derrida argues that the community of world is "always constructed, simulated by a set of stabilizing apparatuses, more or less stable, then, and never natural, language in the broad sense, codes of traces being designed, among all living beings, to construct a unity of world that is always deconstructible, nowhere and never given in nature" (2011, 8-9). In the examples mentioned above, humans and birds find solutions for understanding the other where there is no experience to draw on, and they create meaning by interacting. All beings are thrown into a world of meaning-giving structures that exist, in which new experiences can lead to new meaning. Sometimes this meaning is created in relation to individuals of one's own species; sometimes it is created in interaction with individuals of other species.

### *Baboons and Bonobos*

For twenty-five years, the ethologist Barbara Smuts (2001) studied baboons in Kenya and Tanzania. The baboons she came to know best were the Eburru Cliffs Troop, a group of one hundred and thirty-five baboons who moved in a seventy-square-kilometer territory in search of food. For two years, Smuts travelled with them, from dusk to dawn, for around twelve hours a day. For several months she lived completely without the company of other humans; later on she lived with other researchers, whom she only saw in the evenings and did not have much contact with otherwise. In the beginning of her research, Smuts tried to get as close as she could to the baboons so she could get a better view of their behavior. She approached them until they ran away from her, and then stopped moving, waited until they relaxed and approached them again, but this strategy did not get her very far. After a while she began to see the signals they gave before they started to move away from her—mothers called their children, different members of the group gestured to each other—and learned to stop before she scared them. In this way, she was able to move much closer in a shorter period of time.

When doing research on primates′ behavior in the wild, scientists usually try to ignore them, in order to not let their presence near them influence their interaction. Smuts found out that ignoring them is not a neutral act:

> (A)lthough ignoring the approach of a baboon may at first sound like a good strategy, those who advised me to do so did not take into account the baboons' insistence on regarding me as a social being. After a little while, I stopped reflexively ignoring baboons who approached me and instead varied my response depending on the baboon and the circumstances. Usually, I made brief eye contact or grunted. When I behaved in this baboon-appropriate fashion, the animals generally paid less attention to me than they did if I ignored them. It seemed that they read my signals much as they read each other's. By acknowledging a baboon's presence, I expressed respect, and by responding in ways I picked up from them, I let the baboons know that my intentions were benign and that I assumed they likewise meant me no harm. Once this was clearly communicated in both directions, we could relax in each other's company. (Smuts 2001)

Baboons who are closely related or who are good friends sometimes ignore each other's close presence, but under other circumstances, ignoring someone can express mistrust or tension. Smuts needed to be aware of that and to act accordingly to be able to study the baboons. Furthermore, by interacting with them, she experienced critical aspects of their society, such as hierarchy, personal space, and communication, directly. Because she attended to them and adjusted her behavior, the baboons came to accept her as a social being in their midst; as a subject to communicate with instead of an object that had to be avoided. Scientists call this habituation, which implies that the baboons changed their ways to accept her as a neutral observer, but in Smuts' experience the opposite happened: the baboons went on with their lives while she had to change her ways to be with them (see also Candea 2013; Despret 2008; Haraway 2008).

By living with the baboons, experiencing their habits and daily life, Smuts became attuned to their movements, which changed her experiences of her surroundings. She felt like she was turning into a baboon:

> A simple example involves my reactions to the weather. On the savanna during the rainy season, we could see storms approaching from a great distance. The baboons became restless, anticipating a heavy downpour. At the same time, because they wanted to keep eating, they preferred to stay out in the open as long as possible. The baboons had perfected the art of balancing hunger with the need for shelter. Just when it seemed inevitable to me that we would all get drenched, the

> troop would rise as one and race for the cliffs, reaching protection exactly as big drops began to fall. For many months, I wanted to run well before they did. Then something shifted, and I knew without thinking when it was time to move. I could not attribute this awareness to anything I saw, or heard or smelled; I just knew. Surely it was the same for the baboons. To me, this was a small but significant triumph. (2001)

Smuts emphasizes the physical-material aspects of this process. In his *Nature* Lectures, phenomenologist Maurice Merleau-Ponty introduces the concept "strange kinship" (2003, 271). This kinship allows humans to be together with other embodied beings, not on the basis of a shared origin or evolution, nor language or culture, but because humans have bodies that relate to their environment and other bodies. The relationship between humanity and animality is not hierarchical but lateral. Humanity neither emerges from animality teleologically, nor is it cut off from it. The difference between humanity and animality cannot be reduced to evolution, or to an abyss between body and consciousness, or animal and *Dasein*. According to Oliver (2007), this strange kinship allows for "an intimate relation based on shared embodiment without denying differences between life-styles or styles of being" (Oliver 2007, 18). In *The Visible and the Invisible,* the intertwining of animality and humanity becomes the intertwining of the visible and the invisible, the body and the mind. Thickness of the flesh and permeability of the skin (1964, 141) make "intercorporiety" possible. The thickness of the flesh allows for relations with others, while the skin ensures that we can distinguish our experience from those of others. In communication, our body is always subject and object: we see and are seen, we touch and are touched. Merleau-Ponty calls this "reversibility" (1964, 123). Beings of other species also have flesh and skin, and this transitivity also happens between humans and animals of other species (Oliver 2007).

Scott Churchill (2005) describes encounters with bonobos who live in captivity in Fort Worth Zoo. Drawing on the work of Merleau-Ponty, he argues that an extensive bodily conversation comes into being when he positions himself on the other side of the glass of the baboons and interacts with them by mimicking their movements. He meets with the bonobos often and thinks they appreciate the contact as much as he does. Churchill does not mention the glass between him and the bonobos, nor their captivity. While it is possible that Churchill and the bonobos developed a new language-game, the glass wall between human and bonobos reduces their options for embodied interaction, because the bonobos cannot act in species-specific ways, and because the sensory experience is reduced to sight; the bonobos and Churchill cannot touch or smell the other, and sounds are muted. In conceptualizing interspecies intersubjectivity, Smuts rejects common interpretations of animal subjectivity (see Allen and Bekoff 1999 for an overview), described in terms of self-awareness, consciousness or even social cognition, because these concepts are usually understood as mechanisms that only allow animals to achieve certain instrumental acts. In contrast, Smuts sees something resembling a human self in individuals of different species and she emphasizes their agency. In the situation Churchill describes, human and bonobos create a common language-game, heavily based on mimicry. Because humans decide the framework for macro-agency, possibilities for the bonobos to exercise agency are limited, which limits possibilities to create a common language and world. The setting, a zoo, determines key aspects of the experience and makes it difficult to go beyond a subject-object relationship. In contrast, Smuts really needed to make an effort to communicate with the baboons as subjects, and she needed to change her own behavior to be accepted by them as a social subject. To become part of their group, she needed to pay close attention to the animals she interacted with, and to learn to approach them on their terms. She moved with them for a longer period of time, which enabled her to get an understanding of their language. This changed her experience of her surroundings, and opened possibilities to create a common world with them.

*Being with Other Animals*

In living with the baboons, Smuts kept her distance. She "spoke baboon" and the baboons did their best to understand her, even though she had an "outrageous human accent," they had a mutual understanding. But she did not build close relationships with them, something she did do with her dog companion Safi (Smuts 2001). After she adopted Safi from an animal shelter, Smuts decided not to train Safi, but to communicate with her as an equal: using words, nonverbal vocalizations, body language, gestures, and facial expressions. Safi and Smuts are both sensitive to the other and reach a high degree of understanding and intimacy. Because Safi has an "inherent sense of appropriate behavior in different circumstances," Smuts can take her almost anywhere off leash. In the city, she will make most decisions; outside Safi usually decides where to go and what to do. If they disagree, they meet somewhere in the middle. Smuts describes their daily rituals—including dog-human yoga in the morning, initiated by Safi—that come into being as a result of constant interaction. She sees intersubjectivity in their case as the possibility to shape new realities through interacting with others, in which both subjects are not given before they meet (see also Haraway 2003, 2008). Their habits and rituals (see also Smuts 2006) are not given beforehand but originate from their life together and give it substance.

According to Heidegger, "Being-attuned" to others (1962, 172), *Mitsein,* is a fundamental characteristic of *Dasein,* of the structure of being (human) in the world. Discourse—which is expressed in language and is language—is constitutive for *Dasein*'s existence. Discourse communicates, and this constitutes the articulation of being with one another. We are always with others; this is constitutive for our way of being in the world, which is made explicit in discourse, of which hearing and keeping silent are an important part. The movement is twofold; language shapes our way of being in the world and we shape language. In discourse, humans and other animals bring their memories and histories and relate those to new experiences, thereby creating meaning. The baboons have ways to communicate with one another and for giving meaning to new experiences. Living with Smuts expands their repertoire of meaning-giving structures (i.e., the concept of "friendly stranger" is added). Smuts acts meaningful in a human way, and through discourse comes to understanding with the baboons. Safi and Smuts develop a more complex mutual language through interacting and sharing experiences. Their communication creates a common world, and for them communicating with the other is a fundamental attribute of who they are. Humans and other animals are attuned to one another, and language and discourse are part of creating meaning in their respective lives. Through language, acts gain meaning; in the same movement languages develop through acting with others.

## CONCLUSION

Viewing language as a set of language-games is a good starting point for rethinking language non-anthropocentrically, because it allows for conceptualizing a plurality of ways in which language is used. This is necessary because there is not one "animal" but a multitude of animals who express themselves differently. The concept language-games also challenges the dominant view of language, as having a univocal meaning, as well as the image of the human subject as exceptional and potentially all-knowing. While some forms of language use, such as writing novels, are only found in human languages and cultures, other language-games are also found in other species. Similarly, nonhuman animals of different species have ways of expressing themselves that are unavailable to humans, such as changing patterns on their skin, or using smells that have a grammar. Human language is special, but so is squid language, or bee language. Concepts such as grammar, recursion, or mimicry can help to describe and understand other animals' languages. This does not necessarily mean that these concepts apply in the exact same way to their languages as to human language; nor does it mean that other animals have languages similar to human language. It can, however, help us discover similarities and clarify differences, which can give us a better understanding of the animal language in question, and of the meaning of the concept that is used. In many language-games, verbal and nonverbal acts are intertwined, and in some words play no role at all. The fact that other animals do not use human words does not necessarily imply that their expressions are not language.

In order to further investigate animal languages and the limits of the concept of language, as well as to further develop interspecies understandings, interdisciplinary work is needed. New insights about animal cognition, cultures, and languages change the meaning of our interactions with other animals, and philosophy can help us think this through. On the other hand, philosophy should be informed by progressing insights in biology and ethology, in order not to repeat stereotypes about nonhuman animal behaviors and capacities. As Howard (1952), Lorenz (1949), and Smuts (2001) show, living with other animals provides us with another kind of insight into their lives than scientific studies in the laboratory. New language-games come into being when beings of different species interact, and these can create understanding and open up possibilities for new ways of living together. In the words of Vicki Hearne: "When we learn a language-game, we learn to read the darkness" (1994, 72). While speaking with others—human or nonhuman—does not automatically lead to understanding, the possibility of understanding is there. The outcome is always open.

## NOTES

1. Hereafter abbreviated as *PI*.
2. Recent research shows that the level of oxytocin rises in dogs and their humans when they gaze into each other's eyes (Nagasawa et al. 2015).

# SIX

# Desire and/or Need for Life? Toward a Phenomenological Dialectic of the Organism

*Sebastjan Vörös and Peter Gaitsch*

In recent years, an important change has been underway in biology. After having been dominated by an "unapologetically genocentric view of life" for more than half a century (Nicholson 2014, 348), the shortcomings of the reductionist-mechanist amalgam of Modern Evolutionary Synthesis and molecular biology have caused biologists to question the underpinnings of their discipline and reconsider some central questions that have been relegated to the fringes of legitimate scientific inquiry (Bruce 2014; Denton et al. 2013; Huneman and Wolfe 2010; Pepper and Herron 2014; Wolfe 2010). Although it is perhaps too early to speak of a Kuhnian paradigm shift, the theoretical spasm of "genomic angst" (Denton et al. 2013, 43–44) is now slowly subsiding, and the dawning of the "postgenomic era" (Welch and Clegg 2010) resounds with a loud call for change: "A New Biology for a New Century" (Woese 2004). But what was it that has provoked this claim and what is its central message?

By the mid-twentieth century, as an ongoing struggle between mechanism and organicism had slowly drawn to its close, with the former party declared an indisputable winner, the machine model, that "Ur-metaphor of all modern Science" (Lewontin in Nicholson 2013, 669), had become firmly entrenched in life sciences, prompting a focus shift from the *organismic* to *sub*-organismic (e.g., genes) and *supra*-organismic levels (e.g., populations) (Nicholson 2014, 347). From the newly acquired "worms eye" perspective (Weiss in Denton et al. 2013, 3), the organism became reduced to "a sort of interface between the phenotypic expression of genes and the selecting role of the environment," with no autonomous agency of its own (Nicholson 2014, 348). The big questions, such as "What is life?" and "What is organism?" that had occupied philosophers of biology and theoretical biologists at the end of the nineteenth and the beginning of the twentieth century (cf. Gare 2008, 56–61; Bruce 2014, 356–360) were now deemed obsolete, with the majority of experts tentatively embracing the simple equation: organism and/or life = (gene-driven) machine (Gare 2008, 61; Nicholson 2013, 669).

Recently, however, the conceptual darkness brought on by the "eclipse of the organism" (Nicholson 2014, 356) has been emblazoned by a new wave of criticism directed at the reductionist-mechanist model. Nicholson enumerates three main fronts from which the model has been challenged: (a) *evolutionary biology*: inadequacy of Modern Synthesis to provide a comprehensive account of the evolutionary process; (b) *systems biology*: the failings and limits of unbridled reductionism; and (c) *theoretical biology*: reemergence of the question about the nature of life (Ibid., 348; for an in-depth account see 348–356). In the wake of what some have termed, somewhat polemically, "an organicist revival" (Denton et al. 2013, 31; cf. Nicholson 2014, 356–357), it has thus become obvious that the fundamental questions of biology have never been adequately addressed, but were, for the most part, simply brushed aside and ignored.

In a sense, modern biology is back to square one, for if there is, indeed, to be a "welcome return of the organism" (Bruce 2014, 354), theoretical biologists and philosophers of biology need to "go back to the roots" and confront a vast array of intricate philosophical questions that have lain dormant for more than half a century, questions about the nature of organism, living beings, and life as such. The current chapter purports to provide a small contribution to this overarching agenda by shedding light on these issues from a *phenomenologically inspired* perspective. In other words, it tries to approach the phenomena of "life" and "organism" by drawing from, and reflecting upon, some of the most important phenomenological approaches to biology. But this immediately raises a delicate question: What, if anything, can phenomenology, especially if construed in the Husserlian sense as the study of experience, say about "life" or "organism"? In what sense, and to what degree, do these questions fall under its aegis? Are they not, as matters pertaining to matter (however broadly construed), something that needs to be resolved by natural sciences, and not by reflection on the nature of experience?

## PHENOMENOLOGY OF LIFE? *LEBEN* AND *ERLEBEN*

The French phenomenologist Renaud Barbaras points out that there is a fundamental *tension* in the concept of "life" (Barbaras 2008; 2010; 2012), a tension reflected in the double meaning of the French term *vivre*, designating both "being alive," that is, existence *in* the world (Ger. *leben*), and "the feeling or the experience of something," that is, experience *of* the world (Ger. *erleben*). Life, in its most rudimentary sense, is characterized by a polarity between the "living being" and "lived experience." Now, given that "life presents itself always as organisms" (Bruce 2014, 355), one feels tempted to leave it to biologists to come up with a suitable definition, one anchored in *sheer exteriority*. However, this turn to biology leaves out the second half of the equation:

> In order to work on his object, the biologist must first recognize it, that is, distinguish it, within reality; he must distinguish what is living and what is not. This discrimination is the province of an intuition or an experience that escapes objectification, because an intuition or experience is the condition of its possibility. (Barbaras 2008, 4)

Life as "lived experience" is a condition of possibility for us to be able to identify life as a "living being": it is only because we have the capacity of the former that we are able to recognize the latter. Note, however, that this lived experientiality is not sheer ideality, but is always *embodied*: it is only *in* and *through* the enactment of my corporeality—by *living through* my embodied experience—that I can recognize the contours of a living being: "I cannot understand the function of the living body except by enacting it myself, and except in so far as I am a body which rises toward the world" (Merleau-Ponty 1962, 75). In order to be able to *see* a living being, one first has to *be* one; or as famously stated by Hans Jonas: "[L]ife can only be known by life" (Jonas 1966, 91).

Barbaras argues that, even though the biologist is able to justify his discrimination of what constitutes a living being (e.g., by referring to certain biomolecules), this justification is *always in retrospect*:

> In short, he cannot reintegrate the phenomenal level that gave him access to life into the objective level of his biological analysis, insofar as knowledge presupposes a recognition that is of a different order than itself and that knowledge cannot therefore assimilate. [. . .] Thus, the condition for the possibility of biology is a set of acts of recognition and understanding that take root in my own life inasmuch as I constantly undergo them. (Barbaras 2008, 4–5)

However, this *does not* mean that the reverse is true, that is, that life is *sheer interiority* and can therefore be construed simply as a subjectivity and affectivity. Here, we encounter a symmetrical difficulty to the one faced by the biologist, for it is "undeniable [. . .] that we recognize bodies as living beings precisely within exteriority" and that this is *not* simply my own working—it is not the result of my simply projecting subjectivity and affectivity onto the outside bodies, as Henry (2000) would have it—but rather "*something within exteriority urges me to do that,*" for otherwise I would confer subjectivity/affectivity not only on some, but on *all* external things (Barbaras 2008, 7). Put differently, there has to be a "*mode of pres-*

*ence of living interiority within exteriority,"* for if the latter were absent, it would be impossible for us to identify living beings (Ibid.).

Hence, as Barbaras rightly points out, life needs to be construed as neither exteriority nor interiority, but as "a unique mode of being," as "the original unity of the living being and lived experience" (Barbaras 2008, 7–8). The phenomenon of life is "*neutral* in relation to the division between interior and exterior," it is "a mode of being that is perhaps identically grasped *in me* or *outside of me,* since its immanence contains an exteriorization and its being within the world manifests an interior" (Barbaras 2008, 8; our emphases).

The central aim of this chapter is to take Barbaras' call for a "genuine ontological reform" (Ibid.) seriously, trying to find a way to characterize "life" in a manner that avoids the pitfalls of common metaphysical dualities (subjective-objective, internal-external, etc.). In so doing, it will critically assess some of the most prominent "post-dualist" accounts of life, particularly Jonas' need-driven "ontology of metabolism" and Barbaras' desire-driven "ontology of movement." It will be our contention that both Jonas and Barbaras fall short of achieving this much *needed/desired* goal and that a more integrative view of life, explicated as a *dialectic of need and desire,* is called for. After having outlined the main contours of such a dialectic account, the latter will be used to critically assess the construal of animal life as a model for the living and to examine the presence/absence of desire in plant life.

## NEED, OR HAZARDOUS INDEPENDENCE (JONAS)

The *in-betweenness,* the *neither-exteriority-nor-interiority,* plays a central role in Jonas' philosophy of life. "[T]he organic body," he writes, "signifies the latent crisis of every known ontology" (Jonas 1966, 19), the "crisis" being the result of the fundamental "*bi-unity*" of the living being, of the fact that it designates both "outward" (organism, causality) and "inward form" (selfhood, finality) (Ibid.). The "fact of polarity" (Ibid., 5), the tension between outwardness and inwardness, is thus firmly rooted in the deepest fiber of life and constitutes its essential mode of being. But how are we to account for this bipolar bio-dynamics? What are its origins and its main features?

In what he terms as "an 'existential' interpretation of biological facts" (Ibid., xxiii), Jonas attempts to solve this bio-crisis by paving a way for a new, "post-dualist" monism that would eschew the pitfalls of the Scylla of materialism and the Charybdis of idealism. "Life," he maintains, "does not bear distillation; it is somewhere *between* the[se] purified aspects—*in their concretion.* The abstractions themselves do not live" (Ibid., 28; our emphases). Against these dualities, Jonas proffers a "new reading of the biological record" that might (*pace mind*-less reductive materialism) "recover the inner dimension" and (*pace world*-less idealism) reassert the "psychophysical unity of life" (Ibid., xxiii). However, in order to provide a plausible account of the bi-unity of the living, Jonas must find a way to absorb the polarity between physicalist and phenomenological approaches to life into a "higher unity" (Ibid., 42) and demonstrate that "the organic even in its lowest forms prefigures mind, and that mind even on its farthest reaches remains part of the organic" (Ibid., 1). Put differently, Jonas has to show that only phenomenological aspects of the living are able to identify life in "materiality," while the former are nothing over and above, but are always rooted in, the latter.

How does Jonas set about solving this daunting task? Reflecting on the "ascending scale" in which the variegated life forms present themselves, Jonas claims to have found the "Ariadne's thread" of life (Ibid., 2), the vital *potentia* hidden "in the depths of being" (Ibid., 4). The latter, he feels, is none other than *freedom*—freedom in the sense of a progressively increasing autonomy and complexity of perception and action (Ibid., 2). This organic freedom, however, is haunted by *paradoxicality*: "The privilege of freedom carries the burden of need and means precarious being" (Ibid., 4). Specifically, this "needful freedom" (Ibid., 80) is characterized by "hazardous independence," a perpetual "hovering over t[he existential] abyss" (Ibid.), for in order to preserve its *autonomous outwardness* the living being is condemned to *needful inwardness*. And according to Jonas, the "hidden ground" of this existential interplay between freedom and need, "the original unity" of life in its essential polarity, is *metabolism*:

> In this remarkable mode of being, the material parts of which the organism consists at a given instant are to the penetrating observer only temporary, passing contents whose joint material identity does not coincide with the identity of the whole which they enter and leave, and which sustains its own identity by the very act of foreign matter passing through the system, the living *form*. It is never the same materially and yet persists as its same self, *by* not remaining the same matter. (Ibid., 75–76)

Put differently, metabolic processes enable the organism to constitute itself as an individual identity (stable "organic form") against the constant flux of its temporary material constituents (ever-changing "matter"). Through the process of self-constitution, however, the organism also delimits that which it is not: its "self-isolation" sets it "against the rest of things," against the world (Ibid., 83). To wit, "organic form" constitutes itself against, but is dependent on, "matter."

The fundamental "antinomy of freedom" situated "at the roots of life" (Ibid., 84) thus determines the two fundamental dimensions of its (self-) transcendent nature (Ibid., 85): its *active* aspect designates the outward *reach*, that is, its directedness toward the outside or *exteriority* ("the world"), while its *passive* aspect designates the outward *exposure*, that is, its directedness toward the inside or *interiority* ("the selfhood"). The relation between freedom and necessity, between exteriority ("the world") and interiority ("the self"), is profoundly *dialectical*:

> On the basic level, that defined by metabolism, this double aspect shows in the terms of metabolism itself: denoting, on the side of freedom, a capacity of organic form, namely to change its matter, metabolism denotes equally the irremissible necessity for it to do so. Its 'can' is a 'must' since its execution is identical with being. (Ibid., 83)

In short, the dialectic of exteriority and interiority stems from the dialectic of freedom and necessity; the latter, in turn, is (at least on the fundamental level) enacted in the double aspectivity of metabolism.

Note, however, that metabolism *in itself* is *not* sufficient to *define* life, as it is but one part of the equation. In order to be able to recognize the "organic form" that sets itself apart from the inorganic flux, one has to *be* (*en-act, en-live*) that same form:

> On the strength of the immediate testimony of our bodies we are able to say what no disembodied onlooker would have a cause for saying: that the mathematical God in his homogenous analytical view misses the decisive point—the point of life itself: its self-centred individuality, being for itself and in contraposition to all the rest of the world, with an essential boundary dividing 'inside' and 'outside.' (Ibid., 79)

Only somebody who *embodies* the bi-unity of life is capable of identifying metabolism as its essential vehicle; the *dis-embodied* mathematician-God, despite its alleged omnipotence, is impotent in trying to grasp the vital *potentia* of the living.

Thus, at least *prima facie*, Jonas seems to have succeeded in elucidating life against the background of its fundamental bi-unity. Closer scrutiny, however, shows his dialectical account to be dubious. The most important inconsistency is that the purported dialectic between freedom and need is *posited*, but not corroborated, for even though Jonas *claims* that freedom and need are on the same footing, his metabolism-based explication "*fails to deliver*." As Barbaras points out, Jonas' approach is besmirched by "an excess of interiority" (Barbaras 2010, 95), by what might be called a "hypertrophy of need," and therefore falls short of providing a *positive* account of the dialectical (self-)transcendence. In Jonas, going-beyond-oneself is *not* a manifestation of genuine (indeterminate) openness of the living, but is determined and regulated *by need*: it is, in short, nothing but "the other side of material restoration" (Ibid., 102). The genuinely *self-transcending* momentum (*dynamis*), "an aspiration or tendency that is as yet indeterminate" (Ibid.), is thus thwarted by the *self-maintaining* momentum, the needful gravitation toward *stasis*: life is essentially *survival*, the struggle to *preserve* itself, to *keep* itself alive. It is, in other words, "absolute submission to the pressure of need" (Ibid., 115).

This, Barbaras argues, leads to a *circulus vitiosus* (construing life as survival presupposes what it seeks to explain) and, even more importantly, defines life by its opposite, namely *death* (Ibid.). Moreover, Jonas' account is unable to adequately explain the phenomenon of *movement*: construing movement as the satisfaction of need results in elucidating movement from something other than life itself, from that which puts life in danger (Barbaras 2008, 11). Movement becomes something *extrinsic* to life, something accidental and contingent (Barbaras 2010, 107), and is relegated to the realm of *animality*. From this it follows that, in Jonas, the prime model for living organisms is *plant life*, an animal being only "a plant plus something" (Barbaras 2010, 109), and that, consequently, his account prioritizes the *objective* level of explanation:

> The metabiological, philosophical subordination of the phenomenal order to the objective order is in fact commandeered by an underhand subordination of the phenomenal order to the objective order, of the phenomenological to the biochemical. It is indeed at this objective level that a precise determination of vegetative life is possible. (Ibid., 110)

By overemphasizing necessity, Jonas inadvertently prioritizes *interiority* over exteriority, *vegetative* over animal, *objective* over phenomenological, and thus undermines the postulated dialectic. The inward pull of need restrains the outward push of freedom, limiting it to *mere conservation* and stymieing genuine exploratory tendencies. Despite claims to the opposite, Jonas' explication of the "needful freedom" *is a*symmetrical, with the first element overshadowing, even suffocating, the second one.

## DESIRE, OR CONSTITUTIVE INSATIABILITY (BARBARAS)

Barbaras' attempt to supersede Jonas' account of life grows naturally from his criticism of what he perceives as the latter's major shortcomings. As already seen above, Barbaras is in full agreement with Jonas that, when it comes to explicating the nature of life, a "genuine ontological reform" (Barbaras 2008, 8) is needed, one that would help us move beyond the entrenched dualisms of internality/externality, subject/object, etc. However, he is adamant that (*pace* Jonas) life needs to be approached on the basis *of itself*, and *not* on the basis of *what it is not* (death). In order to achieve the neutralization of the "ontology of death" (2012, 143), a "renewal of the essence of life" (2010, 107) is called for, an elucidation of life in its "original unity" (2008, 8). It was mentioned in the previous section that the "hypertrophy of need" in Jonas' account leads to "an excess of interiority" and the inability to adequately explain movement (Jonas' organism as a *slave to stasis*). This, according to Barbaras, is the fatal flaw in Jonas' attempt to offer a fresh, post-dualist view of life, as it not only misses the (non)point of life, but actually *negates* it: it abolishes that which separates living beings from things. Namely, the sense of being that qualifies a thing is its "static character"—"the thing is what it is"— whereas the sense of being of a living organism is the *very negation of thingness*—it is, first and foremost, *dynamic* (2013, 164).

Life, construed as an "originary" or "primordial dimension" (2012, 139), as a genuine "third kind of being" (distinct from both "objective living being and [subjective] lived experience"; 2008, 12), is thus *movement itself*. On pain of preventing a slide into static thingness, life cannot be construed as "preservation" or "repetition," but rather as "accomplishment" and "creation," as "the very movement by which the living being constitutes itself" (Ibid.). By equating life with preservation, Jonas remains a "prisoner of presuppositions," and therefore overlooks life's "essential mobility" (Ibid., 12). For movement is *not* something that *happens* to a living being (locomotion); it is not its property, but *that by which a being happens*, it is its *essence* (2008, 12; 2010, 112, 108):

> A being can enter into movement empirically only on the condition of being characterized by a fundamental mobility: it has a movement only insofar as it is in some sense movement. [. . .] An essential feature of [organisms] is [. . .] that they are ontologically situated in movement, that they are on the side of mobility, that they are essentially capable of movement. (2010, 106)

This essential movement (as distinguished from the "empirical" movement of the external/material domain) is "ontologically irreducible" (Ibid., 105), and therefore cannot be the result of, but is rather the condition for, the need itself (Ibid., 106): it is what makes the satisfaction of need possible (2008, 11). Moreover, because living beings are essentially movement, they find themselves in constant motion, their mobility (as vividly expressed in play or exploratory activities) exceeding not only the satisfaction of individual needs (Ibid., 14), but even the *very possibility of the (ultimate) satisfaction as such*:

> The distinctive feature of the living being is that nothing appeases its vital tension [...]. The indefinite reopening of vital movements, which could be described in terms of the indeterminate excess of the potential over the actual, reveals a *fundamental incompleteness* at heart of the living being. [. . .] Life's unceasing mobility refers to an *absence* that cannot be filled in and to a *lack of being* that is the very definition of the living being. (Ibid.; our emphases)

Hence, in Barbaras, it is *not need*, but *desire* that ultimately determines the essential movement, the vital *dynamis*: life is characterized by "constitutive insatiability," manifesting itself as unappeasable dis-possession (Ibid., 14, 15), "what calms it at the same time exacerbates it, so that nothing that satisfies it can fulfill it" (2013, 165). And it is only through this "excess of potential" (2008, 14) that genuine self-transcendence becomes possible, something that we saw was lacking in Jonas' need-driven account. In other words, it is only through desire that *true exteriority* can disclose itself. Consequently, the central model for living beings should be sought not in vegetative but in *animal* life: plants are "animals *minus* something," representing the degraded form of desire (2010, 112), whose mode of being is characterized not by a lack of movement, but by *lesser* movement (Ibid., 113).

This, however, opens up a curious correlation between the desire-driven organism and its world: "insatiable advance of desire" is met with "non-positive excess" of the world (2013, 166), for desire never captures its object "except in the mode of the object's own absence," so the world is given it only as what "indefinitely withdraws behind its own appearance" (2012, 142). But this threatens to undermine Barbaras' post-dualist project: if desire perpetually runs up against, but never actually meets, the world, haven't we slid back into a certain form of dualism? Barbaras tries to solve this dilemma by positing an "ontological kinship" between the two elements (Ibid., 145): what makes the relationship between them possible is that they are "made of the same stuff" (Ibid.), that they both belong to a "deeper reality of which they are both modalities and whose nature is a process" (2013, 167). Barbaras identifies this fundamental reality with the "proto-" or "archi-movement" (2012, 146) of nature. However, although archi-movement can account for the *common ground* between the living being and nature, it cannot account for their *distance*: it cannot explain the process of "individuation by separation," which is "the condition of singularity" of organismal existence (2012, 152). The moment of individuation of "movement" (living being) from "archi-movement" (nature) is termed "archi-event," and is construed as the "event of separation," of movement exiling itself into a discrete being (Ibid.).

These two dimensions are said to elucidate the double aspectivity of desire, its being *in* the world (the moment of "community" through archi-*movement*) but not *of* the world (the moment of "split" through archi-*event*) (2013, 171). This has two further consequences. The first is that Barbaras' account calls for a "double surpassing of phenomenology" (2013, 173): the delineation of archi-movement requires us to move towards "ontology" (2012) or "cosmology" (2013), while the delineation of archi-event requires us to go even further, namely to "metaphysics." The second implication is even more radical: it maintains that life *does not coincide* with living beings, but actually *exceeds* them. Moreover, a living being, conceived as a movement which emerges through the separation from archi-movement *via* archi-event, actually *lacks* life (2013, 175). In Barbaras' view, then, a living organism is a being that is *deprived* of archi-life, and manifests its distinct properties (incessant movement, etc.) precisely *due to this fundamental lack*.

Although hearkening to the same post-dualist call, Barbaras thus ends up with a view of life that is very different from that of Jonas. However, in light of the radical conclusions he draws, one wonders whether his portrayal fares any better. The main problem of Barbaras' account seems to be that, unable to move from *anti-* to *syn*-thesis, it entangles itself in predicaments at the other end of the non-dualist spectrum. If Jonas' need-driven conception of life collapses into interiority, then Barbaras' desire-driven life can be said to *disperse into exteriority*; and while Jonas' life has a hard time "getting off the ground," Barbaras' life constantly *overtakes* and eventually even *escapes* itself. In his quest for genuine exteriority, Barbaras seems to have forgotten the other side of the coin: If life is nothing but ceaseless motion, how can one account for the *tendency* towards stability—regardless of how precarious, fragile, or transient—that seems to pervade the realm of the living? If all there is, is the insatiable propulsion of desire, why are there organisms or species and not just a chaotic tumult of activity? In Barbaras' conception of life there is *no* (not even "virtual") *anchoring*, and if there is no anchoring, how can there be movement? Movement *in relation to what*?

It was seen that Barbaras tries to address these inconsistencies by reverting to the ontological/cosmological archi-movement and metaphysical archi-event, but it is not clear whether this actually remedies the situation. Grounding the emergence of the living in archi-event seems nothing short of a *miracle*: not only does it tell us nothing new about the "mystery of origins," but it also leaves the cloud of mystery hanging over the fundamental nature/structure of organismic life. Where Jonas draws on the double-dialectic of metabolism, Barbaras makes no attempt at an explanation and opts for a mere *positum* instead: living being, construed as a process, is a negation of nature, construed as an Ur-Process. But why and how is this self-exodus of desire enacted (not to mention that a *self*-exodus already *presupposes* a (self-)reference point)? And given the initial motivation for Barbaras' undertaking—to develop a coherent phenomenologically inspired post-dualist conception of life—it is far from obvious why Barbaras' cosmological-metaphysical account should fare better than (m)any of the already existing *metaphysical* solution(s). Although Barbaras insists that the double surpassing of phenomenology remains grounded in phenomenological findings, the end result—the baroque speculative edifice of archi-flux—leaves one wondering.

## DESIRE-NEED, OR VITAL BETWIXT

The end result of our (perhaps somewhat oversimplified) analysis seems to be a curiously symmetrical breakdown of the dialectic: by succumbing to *centripetal* forces, Jonas ends up with *static interiority*, and by succumbing to *centrifugal* forces, Barbaras ends up with *dynamic exteriority*. It turns out, however, that this "symmetrical asymmetry" has telling consequences that are worth spelling out in detail. In Jonas' need-driven account, the dialectical process solidifies into a "rigid point," that is, individuality is construed as something fixed, as something that is prefigured and simply needs to be preserved. However, we have seen that self-preservation undermines self-transcendence and is, as Goldstein points out, ultimately a *pathological* phenomenon: "The tendency to maintain the existent state [. . .] is a sign of anomalous life, of decay of life. The tendency of normal life is toward activity and progress" (Goldstein 1995, 162). In Barbaras' desire-driven account, on the other hand, the dialectical process disperses into an "amorphous flux," that is, individuality is never reached, it dissolves in perpetual change. However, we have seen that motility also requires "anchoring" or, as Plessner is keen to emphasize, *delimitation*: the fact of setting and handling its own limits: "This feature, which can be found in all organisms in all modes of organization, is to be accounted for as a minimum condition, according to which, insofar as it is fulfilled, constitutes life [*Lebendigkeit*]" (Plessner 1975, xxx). Ultimately, the breakdown in the purported dialectic seems to have resulted in a complementary one-sidedness. We could even exaggerate this point somewhat by saying: Jonas needs Barbaras' desire (exteriority), while Barbaras desires Jonas' need (interiority). What, at any rate, is the reason for this fearful *a*symmetry, and how, if at all, could it be sidestepped?

Interestingly, where Jonas and Barbaras meet is in that they both construe life as *lack*[1]: the posited vital dialectic—the dynamic of needful freedom in Jonas and the tension between archi-movement and archi-event in Barbaras—enacts a certain *absence*, a certain *already-beyond*, a certain *not-yet-there*. But as soon as they try to *conceptualize* this lack, it breaks down into a one-sided (metaphysical) system incapable of sustaining the postulated dialectic. The problem, we feel, lies at the very root of their approach, in the fact that, in the last analysis, they both try to subsume life under an (admittedly non-dualist) common metaphysical denominator. However, in order to take the groundlessness of life *seriously*, it is not enough to strive only for a *post-dualist*, but also for a *post-monist* account of life. In other words, if the posited vital dialectic is to "take root," it is important to wholeheartedly confront the *horror vacui*, that old spectre of Western metaphysics. Jonas and Barbaras seem to fall prey to the same song of sirens, touching upon life's groundlessness, but then immediately moving on to something more substantial, something (metaphysically) more palatable.

The lack that is life is thus not sheer nullity (substantialized nothingness), but is enacted precisely in-and-through this fundamental circularity of the vital dialectic. It is the *in-betweenness*, the *vital betwixt* between the inward- and the outward-movement, between self-subsistence and self-transcendence. And this "pulsating polarity" manifests itself in both aspects of life, in living beings and in lived experience. It is, indeed, this conceptually elusive dialectical betwixt that enables convergence of the two perspectives, constituting the "groundless ground" wherein they meet.

On the side of the "living being," it was argued that the organism must not be dominated by either need or desire, but can be said to exist only against the background of "creative tension" between them. It is, so to speak, *constitutive self-transcendence*: it constitutes itself in self-transcendence, and transcends itself in self-constitution: it never fully captures itself, yet it also never fully escapes itself. Here, the chronically understudied Plessner can serve as a wellspring of inspiration. Following his theory of positionality, life must (*pace* Barbaras) not only flow, but also (by constituting an organismic system) *retain*; it must (*pace* Jonas) not only stay-in-itself, but also (by investing itself in an incalculable developmental process) *become-the-other*. The living being is thus fundamentally characterized by a "double aspectivity of two opposite directions not convertible into one another" (Plessner 1982, 9), by a "double transcendence" of interiorization and exteriorization (Plessner 1975, 130). By *contraction*, it withholds itself prior to its limits, yet by *expansion*, it concomitantly moves beyond those limits. Barbaras is therefore right in maintaining that the organism is not the same as life, but his claim needs to be qualified by stating that the organism is *traversed* by life.

The same dialectical complementarity that is at work in the living organism can also be found on the side of "lived experience." In his illuminating study of the lived body, Leder notes the paradoxical nature of the incarnality of experience reflected in the fact that our *bodily presence* seldom becomes the thematic object of experience and is therefore essentially characterized by *absence* (Leder 1990, 1). My lived body is ultimately what Husserl termed as *Nullpunkt* (nullpoint), an absent, lived "here" around which all present "theres" are arranged (Ibid., 13). Further, it is characterized by *double* "awayness" (Ibid., 69): the *ecstatic* body (bodily surface, "flesh") constantly *moves-beyond*-itself (projects outward), while the *recessive* body (inner body, "blood") *withdraws-within*-itself (recedes inward) (Ibid., 56). This embodied nullity is thus lived out as "flesh and blood," as an ecstatic/recessive being-in-the world,

> Engaged both in leaping out and falling back. Through its sensorimotor surface it projects outward to the world. At the same time it recedes from its own apprehension into anonymous visceral depths. The body is never a simple presence, but that which is away from itself, a being of difference and absence. (Ibid., 103)

Reverting to the initial Jonasian postulate that life can only be known by life, we can further qualify this statement by saying that it is only by embodying the vital in-betweenness of need and desire that one is able to recognize it enacted in the living organisms. It is this existential version of "deep calling unto deep," groundlessness calling unto groundlessness, that lies at the heart of the dialectical bio-unity of the living. And as soon as the dialectical dynamic is broken, the (non)point of life is negated. Just as lived body ceases to be lived at moments of breakdowns, that is, when the ecstatic/recessive move of *dis*-appearance breaks down into the static mode of *dys*-appearance (disease, suffering, fatigue, etc.) (Ibid., 83–84), so the non-dualist dialectic ceases to be dialectical when exposed to the disembodied (reflective) gaze. It is the pained dis-ease of the objectifying/theorizing gaze that causes the dialectical "dis" to lapse into a solidified "dys" and thereby transform the *absence* of *null*point into the *presence* of *full*point.

To recapitulate: The post-dualist conception of life, it will be remembered, calls for a "genuine [post-dualist] ontological reform," (re)construing life as a "unique mode of being." However, it was shown that, while both Jonas and Barbaras are correct in trying to explicate life against a dynamic interplay between internality-externality, their quest for a post-dualist *monism* ultimately causes them to miss their target. They *start* from the dialectical "tension" of life, but as soon as they touch upon its groundlessness, they set out looking for a higher (monist) synthesis and thus end up *explaining it away*. It is our contention, however, that the "vital groundlessness" must not be shunned but wholeheartedly embraced: instead of trying to *translate* it into something metaphysically more palatable, it needs to be vigorously affirmed. Hence, a genuine ontological reform calls not for a monist, but an explicitly *non-dualist* account of life, an account that tries to elucidate life in terms of *dialectical in-betweenness* or *vital betwixt*, embodied in and through the inward pull of need and outward push of desire. Of course, one could argue that our cursory treatment of the topic leaves much to be desired—and rightfully so—but it should be noted that our intention was not so much to provide a full-blown explanation, but rather a *preliminary exposition*, one that argues for the need of (re)introducing (decidedly affirming) concepts such as in-betweenness, groundlessness, and non-duality when addressing the problem of life. To see how this approach might be tentatively used in a more concrete setting, let us take a brief look at the notion of "lack" in the context of plant life.

## ANIMALISM, AND WHAT LACKS: PLANT LIFE

At first glance, it might seem surprising to label our provisional result, an ostensibly speculative view of life grounded in the groundless dialectic between need and desire, "phenomenological." But it should be clear by now that the vindication for our mediating approach is founded on the fact that life cannot be exhausted by a list of immediately accessible phenomenological features—what Plessner calls "indicative essential features," for example, plasticity of form or rhythmic movement—but calls for a more profound conceptual strategy, which would enable us to hone in on its evasive "constitutive essential features" (Plessner 1975, 114). In this sense, the present section is meant to show that our dialectical "formula" can indeed be applied as a most informative interpretative key for elucidating characteristic differences and commonalities in animal and plant life.

The first problem arises when we think of the diversity and polymorphism of living beings. Is our dialectical approach actually capable of comprehending life in its multiplicity? As we have seen, Barbaras' desire-driven account implies that "animals [. . .] characterized by their mobility are the prototype of living organisms" (Barbaras 2010, 111), whereas Jonas' need-driven account implies that plant life is the prime model of living organisms (Ibid., 109). If our balancing model is correct, it should elucidate not only the need-driven behaviors in animals, but also the *desire-driven movements in plants*—provided, of course, there are any. The pressing doubts about this latter point indicate that the true "litmus test" for the dialectical characterization of life is the *non*-animal, or in our case, *plant* life.

The view that animals, defined by cognitive ethology as "essentially 'subjectively centered' living creatures," are the prime model of living beings is called "animalism" (Hanna and Thompson 2003) or "zoocentrism" (Marder 2014) and arguably represents the main current in philosophical biology. For if life can only be known by life (Jonas 1966, 91), it is my own *animal* living body which serves as a model for every possible living body that we might come across as an object. This "top-down" approach to life has been familiar to phenomenology ever since Husserl first suggested that all knowledge of life is built on our empathic directedness toward other living beings grounded in our own human corporeality (Husserl 1993, 329). However, a reasonable suspicion is that plant life stands at the ineradicable "limits of empathy" (Marder 2012a), constituting the "vast green blind spot" (Marder 2014) of all phenomenological top-down approaches to life. Symptomatically, a recent attempt at "biophenomenology," aiming at "a phenomenological description of the livingness of life" (Sachs and Epting 2010, 14), does not devote a single word to plant life. Moreover, the specifics of plant life are not only neglected by zoocentrism/animalism, but also by the ecological reevaluation of all living things in nature labeled "biocentrism," which turns life into an "amorphous category": "It is as though the life of plants slips between the cracks of *bios* and *zoe* when it is considered, at best, to be the deficient modality of animal existence" (Marder 2014).

The glaring shortcomings of animalism should lead us to consider "phytocentrism," a position that shifts the focus to the fact that, like plants, all living beings are—*phuomena*—a community of growing things (Ibid.).[2] Put differently, there exists a "bottom-up" or "synecdochic logic" of vegetal life (Ibid.): plants are instances that can stand for the whole of life. This is not so say that animalism is completely wrong; rather, animalism retains its role as a methodologically indispensable instrument for approaching life, but it needs to be balanced by a bottom-up approach, such as phytocentrism. This, in line with our previous suggestions, would set up a productive circulatory motion dynamically interconnecting top-down (grounded in empathic "lived experience") and bottom-up approaches (grounded in "living organism").

Phytocentrism's focus on plants resonates with new trends in plant sciences, which seem to be uncovering more and more examples of "intelligent behavior" in plants (cf. Chamovitz 2012; Pollan 2013). This has led to the emergence of "plant neurobiology," a controversial new organicist field within plant sciences with potentially revolutionary implications for life sciences in general (Baluška and Mancuso 2007, 205). However, it should be noted that, although this new area of research "entails breaking down the walls between the kingdoms of plants and animals" (Pollan 2013) and harbors a critical attitude toward anthropocentric and cerebrocentric definitions of intelligent behavior, it is nevertheless profoundly marked by animalism, not only in its methods, but also in its general view on life. In other words, there exists a *revealing antagonism* within the new research field: On the one hand, it speaks of the "animal-like capabilities" of plants, and in the most extreme case conceives of the plant as an "upside-down animal, with its main sensory organs and 'brain' on the bottom, underground, and its sexual organs on top" (Ibid.). On the other hand, drawing on the distributed, decentralized, and modular character of plant life, the inconsistencies of animalistic approaches are the target of severe criticism: "Some scientists working on plant intelligence have questioned whether the 'animal-centric' emphasis, along with the obsession with the term 'neurobiology,' has been a mistake and possibly an insult to the plants." (Ibid.).

In order to fend off animalist tendencies in plant sciences, phytocentrism explicitly refuses to make the plant into an alternative center and actually *dismisses centrism as such*: "Phytocentrism does not reconstitute a more authentic center of existence; to the contrary, it names the immanent implosion of this center, without being seduced by the fantasy of a purely fragmentary nature of existence" (Marder 2014). Phyto-*de*-centrism thus places the plant in the center, precisely because it is, in itself, *de-centered*: "The plant lacks itself a vital center, equivalent to the heart or the brain of an animal" (Ibid.). This recognition of a *lack* constitutive of plant life brings us to the crucial question concerning its status: is it only an epistemic denial, a *negation* of a trait possessed by all animals, or is it indeed a trace of *existential negativity*, of "non-being [. . .] at the heart of [. . .] being" (Barbaras 2010, 118)?

According to Jonas' need-driven account, this lack could only signify the absence of a lack, that is, a lack of a (genuinely existential) lack (Jonas 1966, 104). The reason is that, in Jonas, the plant entertains an *immediate* relation to the world, whereas the abilities that are characteristic of the animal—motility, emotion, and perception—rest on the "principle of mediacy" or "distance" (Ibid., 102). The "great secret of animal life" is said to reside in the "loss of immediacy," in "the gap which it is able to maintain between immediate concern and immediate satisfaction" (Ibid.). The plant, however, lacks precisely this distance, for its (temporally) immediate satisfaction by (spatially) immediate contact leaves no room for an upcoming desire: "In this condition of continuous feeding there is no room for desire. Need passes of itself over into satisfaction by the steady operation of the metabolic dynamics" (Ibid.).

The straightforward empirical retort is that recent research on plant intelligence has uncovered many plant behaviors that directly contradict this view of immediacy, including the plants' use of *Fernsinne*-like sight and sound perception (cf. Pollan 2013). But it is crucial to note that, with this refutation, a conceptual realignment is also needed. As pointed out by Barbaras, desire is not (*pace* Jonas) only "modified need," that is, the product of "need" *plus* "distance" (Barbaras 2010, 108). But this does not mean that we, like Barbaras, should simply reverse the whole relation of need and desire by defining the former in terms of the latter. However, to account for the essential mobility of plants, their sessile lifestyle notwithstanding, it seems to be necessary to distinguish between need and desire as *two qualitatively distinct existential modes*. Put differently, desire is not born out of unsatisfied needs brought about by distance, but has its own roots. This is vividly expressed by the fact that plants manifest "inchoate" (nascent) movements *despite* proximity (Ibid., 113).

Indeed, as Marder indicates, recent research in plant intelligence shows that plants engage not only in *restorative*, but also in *explorative* movement: "Although they appear to be anchored in a place, plants incessantly explore their environments, maximizing their exposure to sunlight, avoiding or growing towards the roots of their neighbours and monitoring and responding to changing environmental conditions" (Marder 2012b, 1367). Such findings demonstrate that, in addition to need-driven movements, plants also exhibit desire-driven movements and that, consequently, the dynamic polarity of need and desire is operative within all vital phenomena.

Note, however, that when speaking of plants, we are dealing with a *minimal* form of desire. Perhaps the most important lesson to be learned from all this is that not only does the philosophical quest for life require a post-dualist *and* post-monist solution, but also that *the post-monist lack comes in degrees*. For if vital activity is "movement of individuation" (Barbaras 2010, 120), and if there is no particular movement of individuation manifested in plants, should we then conclude that there is no vitality whatsoever in plants? Instead of accepting such an untimely conclusion, we may—in a meshwork of animalism and phytocentrism—conceive of plants as a "minimal modality" of animals (Ibid., 112). Barbaras has already indicated differences in degree within the "vital lack" by distinguishing the "exile" character of human life from the "exodus" character of animal life, referring to the fact that, within the vital dynamic of cosmological archi-movement and metaphysical archi-event, the human being is *more* of a "metaphysical being" and the animal *more* of a "cosmic creature" (Barbaras 2012, 110). We simply wish to stress that the plant needs to be integrated more comprehensively into the picture, especially since it might turn out to be the originatory non-point of desire. On the one hand, the indication of a minimal lack and therefore a minimal mode of desire should not (*pace* bottom-up approaches, e.g., Jonas) tacitly be read as the denial of desire as such; but, on the other hand, to acknowledge the presence of desire in plants must not (*pace* top-down approaches, e.g., Barbaras) lead to the subordination of the need-driven motility inspired by animalism.

To conclude, the focus on plant life urges us to establish what Plessner termed as a "psychophysical neutral language" (Plessner 1975, 244) that would encompass all life forms and enable us to construe life in terms of vital in-betweenness. To cope phenomenologically with the evasive biunity of life we thus had to break through the objective as well as the phenomenal level and turn squarely to structural and dialectical issues. This enabled us to say that, in plant life, the polarity of need and desire is minimally existent insofar as it is *structurally present* (in the mode of an "inapparent" (*unscheinbar*) low-level dialectic (de)centered around a vital lack)—but also (to the best of our current knowledge) *phenomenally vacant*.

## NOTES

1. For an interesting and somewhat different "absentist" approach to life, see Deacon 2011a, b.
2. For a more comprehensive picture of phytocentrism in the sense of the ancient idea of a plant soul, see Ingensiep 2001; for more recent work on plant semiosis, see Affifi 2013, Krampen 2010, and Kull 2000.

*III*

# Beast No More

# SEVEN

# Understanding the Meaning of Wolf Resurgence, Ecosemiotics, and Landscape Hermeneutics

*Martin Drenthen*

We have entered the era of the Anthropocene, so it goes. More and more people are becoming aware that the planet we live on is increasingly changed by human activity. When in 1980, Caroline Merchant (1980) talked about the "death of nature," and criticized the death of the *concept* of nature, she did not foresee nature itself disappearing. In 1989, conservationist Bill McKibben (1989), in one of the first public books[1] on global warming, proclaimed "the death of nature" and stressed that nature no longer was a force independent of humans but instead was deeply influenced by humans. This caused a shock in the public debate. In 2000, Nobel Prize Laureate Paul Crutzen and his colleague Eugene Stoermer proposed using the term Anthropocene for the current geological epoch—which was then still considered a controversial position (Crutzen & Stoermer 2000).

Today, saying that we have entered the Anthropocene almost feels like a truism. An increasing number of people are coming to terms with the fact that humans are the major force of environmental change on the planet. What is more, whereas earlier in the debate on the Anthropocene the context was that of a threatening global ecological crisis, today the tone of the debate seems to have shifted. Although today many still stress the challenges that the new era presents, others are arguing that the end of nature can be a liberating experience that offers new opportunities. The authors of the Ecomodernist manifesto (Asafu-Adjaye et al., 2015) "reject" the idea "that human societies must harmonize with nature to avoid economic and ecological collapse" and instead rely on technologies to help us "decouple human development" from ecology. They argue that it is up to humankind to decide which parts of nature are useful to us and save those ecosystem services that we consider worthwhile protecting. According to Mark Lynas, author of the book with the telling title *The God Species* (Lynas 2011) and one of the co-authors of the manifesto, "the essential truth of the Anthropocene is this: neither God nor Gaia is in charge. We are. We now get to decide everything from the pH of the oceans to the temperature of the biosphere to the very composition and future evolutionary path of life on Earth" (Lynas 2015).

The ecomodernists clearly advocate the idea that humans are in charge that is: those humans not only need to, but can and should control nature by means of technology and economic valorization. I argue, though, that ecomodernism is based on an utterly flawed assumption, namely that the increased human influence on the planet leads to an increased control of nature. Whereas ecomodernists set out to decouple the human and the natural world, in fact both worlds get increasingly intertwined. Ecomodernists fail to see that increased human influence makes the planet less predictable and controllable. Moreover, they fail to acknowledge that humans are not the only agents in the world of the Anthropocene, there is nonhuman agency too.

When McKibben recognized that pristine, untouched nature hardly exists anymore, he did lay out two possible responses to this situation: the first, "the defiant reflex," was to develop better ways of managing the world; the "more humble" response would be that humankind would learn that it doesn't need to be in control. I believe that this second option is much more appropriate in light of what we know about our current ecological condition. It may be true that we live in a world that is increasingly changed by human activity, but we humans are not the only agents in the world of the Anthropocene (although it may be that we are the only species able to recognize that fact). We still share that world with other species, species that have their own agency. This becomes clear in the case of resurging wildlife in Europe.

The European landscape is rapidly changing due to human impacts. Humans have transformed the European landscape for many centuries, by farming, building infrastructure and cities: cultivating the land and domesticating other species. More recently, humans are having another kind of impact: in an effort to compensate for centuries of environmental decline, national governments and NGOs attempt to increase the size of natural areas in Europe and connect them (Coleman & Aykroyd 2009). The concept of rewilding is becoming more influential in European nature conservation. Most often, the term refers to the return of habitats to a "natural state," or to "the mass restoration of ecosystems" (Monbiot 2013), or to the deliberate release of "missing" species into the wild. For others, rewilding refers to "action at the landscape level with the goal of reducing human control and allowing ecological and evolutionary processes to reassert themselves" (Klyza 2001, 285). Some use the term for the reversal of human "domestication."

Rewilding projects typically start with the designation of new, large-scale habitat areas, and the reintroduction of extirpated species. The establishment of large-scale wilderness areas is meant to create stable refuges for biodiversity, whereas the European ecological network Natura 2000 connects existing natural areas so that species can migrate more easily and biodiversity loss due to fragmentation is counteracted. But although rewilding starts with human actions and policies, these only form the starting point. Rewilding implies an invitation to non-human nature to respond to our initial actions. Rewilding can therefore better be thought of as the attempt to re-open the communication between humans and non-human species and to acknowledge the need to renegotiate the use of shared space between them.

But wild nature does not just result from human nature policies; it also resurfaces spontaneously (Höchtl et al. 2005; Hunziker 1995). As the European human population decreases and people are moving to the urban centers, rural regions get abandoned. As soon as the human influence on the landscape becomes less dominant, non-human species take the opportunity to occupy new habitats. But the resurgence of wild species is not limited to rural, fairly uninhabited zones. In some urban zones, too, urban adapters such as fox and stone marten increasingly roam the city centers and suburbs. Lynx, bison, beaver, and wild boar are already repopulating areas they had disappeared from centuries ago. In Europe, the number of beavers showed an 14,055 percent increase since 1960, European bison 3,084 percent since 1960, Grey seal 893 percent since 1977, Pyrenean ibex 855 percent since 1960, Southern chamois 537 percent since 1970 (Deinet et al. 2013). A recent study found that today Europe has twice as many wolves as the United States, despite its being half the size and more than twice as densely populated (Chapron et al. 2014). Large carnivores are re-populating Europe's humanly dominated landscapes. Apparently, Europeans somehow manage to coexist with species that recently were hunted down and extirpated. Although rewilding projects occasionally meet local resistance, particularly in areas with a long cultural history, mostly they are applauded by the general public.

The changes to the European landscape have had several effects on animal behavior. These changes are meticulously studied: ecologists and natural scientists study the changing patterns of animal behavior and take note of shifts in habitat distribution and food webs. In response to the omnipresence of humans, wild animals get more and more used to humans and change their behavior accordingly. Sociologists, anthropologists, and cultural geographers examine shifts in human perceptions and use of landscapes, and study changes in power relationships on the land as a result of landscape change. The animal world as well as the human world is changing, and both changes are related.

One of the most spectacular examples of these landscape changes is the return of the wolf (*Canis lupus lupus*) in Europe. Wolves are returning to landscapes they had been absent from for many decades; they are now recolonizing parts of the European landscape that not just bear the marks of human presence, but that are thoroughly humanized. Wolves are discovering how to live close to human civilization. Conversely, humans living in these cultural landscapes have to learn what it means to live in increasingly *hybrid* landscapes, in which wild and tame get intertwined.

Contrary to widespread traditional ideas about the relation between culture and nature, there is no clear distinction between humanly dominated, domesticated cultural land and untouched, pristine natural wildernesses. In the long debates about the Anthropocene, many authors have convincingly shown that humans have had an extensive impact on landscapes for ages already, even long before today's agricultural practice.[2] Yet, recently, we seem to witness a proliferation of hybrid landscapes, as more non-human species are recolonizing humanly dominated territory. This process does not only have material aspects, but can also be studied in terms of semiotic communication. As humans and nonhumans live together ever more closely, their respective spheres of significance are becoming increasingly entangled too. Changes in the landscape change the relation between humans and non-human animals, and both nonhumans and humans need to respond to the new situation. For humans, these changes may challenge existing notions of moral environmental identity, among other things, and call for a revision of existing understandings of the world they inhabit and of the creatures that they share their world with.

## ECOSEMIOTIC APPROACH TO LANDSCAPE CHANGE

An ecosemiotic analysis may contribute to a better understanding of changing landscapes and changing human-animal relationships. An example of such an approach can be found in Timo Maran's article about the appearance of the golden jackals (*Canis aureus*) in Estonia (2015).[3] Maran sums up nicely what is the methodological starting point of an ecosemiotics approach to species management:

1. Natural processes, nonhuman animals and human social groups are all considered as agents that are able to initiate change and that can be influenced by the change initiated by other parties.
2. All living parties (species, different social groups) are considered to be semiotic subjects. Semiotic subjects seek to understand their environment. In doing this, semiotic subjects have abilities of modeling (that is, making sense of surrounding environments, process-

es, and actors on the basis of analogical relations and self-modeling).
3. The process of species management is seen as a dynamical interaction of self-modeling and modeling processes between different human parties and biological species. In this process, different means of communication, translation, and persuasion are used. (Maran 2015, 3)

Maran analyzes the development of the discourse about the appearance of a new species, the golden jackals, in the Estonian landscape, with the aim of helping species management and communication about human-animal interactions to better deal with changes in the "encounter of different semiotic subjects"(2015, 3). Maran shows how the discourse evolved when the new species entered the Estonian landscape, and reflects shifts in the interactions between the jackal and humans. It appears that the golden jackal's semiotic activity played an essential role in the development of the discourse, but at the same time, he notes that "human cultural models also influence the interpretation of a new species to a considerable degree" (2015,1). In order "to improve human communication about new or invasive species," Maran suggests "to raise awareness of the underlying cultural models and to use integrative communication as the developing discourse is dynamical and constantly changing for all interest groups" (2015,1).

A semiotic analysis like the one presented here can contribute significantly to a better understanding of changing animal-human relationships. It can help explain the interrelation between the way that animals make sense of their world and the way that people think of these animals, and point to changes in the semiosphere due to recent environmental changes. A semiotic analysis may even explain how certain cultural ideas about animals and their place in the landscape have become problematic in light of recent environmental changes. Yet, it can be questioned whether this semiotic explanation alone is enough to arrive at an adequate normative assessment of the situation. The strength of an ecosemiotic approach lies in its ability to make a relational analysis of the semiosphere, but its limitation lies as well in its "explanatory" form of understanding. And although the ecosemiotic analysis does include the first-person perspective of both human and non-human beings, the very "impartiality" of its approach also inhibits an adequate understanding of the normative issues at stake for us humans.

There is a relevant difference between the kind of semiotic understanding of signs that humans share with other animals, and the typical hermeneutic interpretative understanding of meanings that is typical for humans.[4] Animals understand the world as a correlate of their sensory apparatus; they understand functional relationships between their own sensory existence, and their surroundings, and thus form a representation or model of the world. Conversely, their communication forms consist of exchanging signs that represent aspects of their relationship to their environment. Human understanding of meaning, in contrast, transcends this mere "instrumental" relationship. Human interpretations of the world do not so much *re*present, but rather *present* a world; and thus they transform a simple environment ("Umwelt") into a world "that one could inhabit," to use a phrase by Paul Ricoeur (Ricoeur 1991, p.149).

The question of moral meaning necessarily presupposes a human perspective, as Bernard Williams has convincingly argued.[5] This does not mean, of course, that the non-human perspective on the world is irrelevant, morally or otherwise. It simply means that the question of the moral significance and meaning of a specific situation is a question that is addressed to us humans, as members of a human language community embedded in a cultural tradition. Taking seriously the moral dimension of a situation requires that we take that specific human understanding of the world seriously. The question of why and how *non*human perspectives on the world should matter to us, is itself a thoroughly human question of meaning.

Environmental hermeneutics is the approach in environmental philosophy and ethics that seek to do justice to the central role of interpretation in the (normative) relationship between humans and their environment. According to hermenutics, all understanding is self-understanding. From a hermeneutic perspective, the very idea that meanings can exist outside the realm of human understanding and interpretation is by itself incomprehensible. Meaning is tied to the perspective of historical beings that are "suspended in language," beings that are capable of understanding meanings.

The changing role of animals in hybrid landscapes not only raises questions that demand an explanation, but can also appear to us as a meaningful situation that beckons to be understood. The answer to the question of what the meaning of the resurgence of wild wolves in our landscapes is, and how one should respond to it, is a question that is as much about the wolf as it is about ourselves. Therefore, a full understanding of the meaning of changing landscapes and the resurgence of predators requires that the semiotic explanatory approach is complemented by another, more interpretative, hermeneutic approach.

In this chapter, I show how the work of Paul Ricoeur can help shed light on the relation between an ecosemiotic and a hermeneutic approach to landscape change. Ricoeur distinguishes between two approaches to reading texts: a "structural analysis" of a text that seeks to understand the internal structure of a text by examining the relationship of textual elements to each other, and a hermeneutic interpretation of the text, that begins with the dialectic between reader and text, and seeks to understand the text as "a world that one might inhabit." I will first present

Ricoeur's distinction a bit more in detail, and will then show how it can help think through how a semiotic analysis of wolf resurgence can and must be complemented by a hermeneutic inquiry into the meaning of the return of the wolves.

## RICOEUR ON READING TEXTS

In his essay "What is a text? Explanation and understanding" (Ricoeur 1981), Ricoeur sets out to show that explanation and interpretation are not opposite modes of understanding, but rather complementary ways of approaching a text, both of which are legitimate responses to the nature of a text. What sets a text apart from other forms of language use, is that a text is "a discourse fixed by writing" (Ricoeur 1981, 146). Ricoeur explains what a text is by comparing written texts with the use of language in speech. According to Ricoeur the significant difference between texts and speech is that in speech, a speaker can accompany his signs and explain himself, whereas texts (and other text-like "things") are not self-explanatory and therefore require interpretation. When language is transformed into a *text*, Ricoeur argues, it assumes a life of its own, or to use a typical Ricoeurian phrase: "the text has emancipated from its author."[6] Without an external authoritative source to turn to regarding the meaning of a text, a reader can only revert to the act of reading the text itself. Moreover, a text is also emancipated from the world; in a text the relation with the world is also "interrupted." Whereas in speech, a speaker can literally point to the things he is talking about; a speaker presents to an interlocutor a "real" world of which both speaker and interlocutor are part, a text brings forth a world of its own, an imaginary world that is a closed structure of references. The reader is needed for that text to become "alive" again, if only because gaps in the text's internal references must ultimately be filled by the imagination of the reader in order to make sense.

Because of this emancipation from author and world, a text can be approached in two distinct ways:

> We can, as readers, remain in the suspense of the text, treating it as a worldless and authorless object; in this case, we explain the text in terms of its internal relations, its structure. On the other hand, we can lift the suspense and fulfill the text in speech, restoring it to living communication; in this case, we interpret the text. These two possibilities both belong to reading, and reading is the dialectic of these two attitudes. (Ricoeur 1981, 152)

Explanation and interpretation are therefore not opposite modes of understanding, but rather different, complementary ways of reading a text that both have their own role to play in the act of reading. When linguists and other structuralists set out to *explain* how a text "works," how it is internally structured, and how the different language elements in a text are related, they focus on a "structural analysis," their text "has no outside but only an inside; it has no transcendent aim, unlike a speech that is addressed to someone about something"—a kind of reading that "is not only possible but legitimate" (Ricoeur 1981, 153). The second approach to reading, in contrast, does not seek to explain how a text is internally structured, but rather asks what the text has to say about the world, what its meaning is *for the reader*. According to Ricoeur, "it is this second attitude that is the real aim of reading. [. . .] If reading is possible, it is indeed because the text is not closed in on itself but opens out onto other things" (Ricoeur 1981, 158). Only in this second approach can "texts speak of possible worlds and of possible ways of orienting oneself in these worlds" (Ricoeur 1981, 177). If the reader answers to the "invitation of the text," then the "refiguration of the world by the text" can bring about an active reorganization of the reader's being-in-the-world. It is through the texts one reads and by imagining oneself in the meaningful worlds that are being opened by these texts that one gets to know "oneself as another" (Ricoeur 1992).

Understanding the meaning of a text thus requires an openness to the world as presented by the text, and a willingness to "place oneself"—for the time being—in that world: text, world, and reader are engaged in a dialectical relationship. Yet, understanding meanings requires an active appropriation by the reader. Texts may appeal to us with a claim to significance, but their *meaning* only becomes clear once we attempt to interpretatively appropriate or "bring home" what it is that beckons to be understood.[7]

In other words: good reading requires an openness for the "strangeness" of a text ("*distantiation*") and a willingness to abstract from the context of one's particular life. But it also demands involvement, active participation, and a willingness to bring to bear the meanings of words and concepts that play a role in his own life ("*appropriation*"). Ricoeur argues that both modes of reading are intertwined and presuppose each other: explanation implicitly rests on a background hermeneutic understanding of the text as somehow worth while analyzing. At the same time, a serious hermeneutic interpretation cannot do without a systematic process of explanation, because it brings into play the "objectivity" of a text, conveying certain structural features of the text that need to be acknowledged if an interpretative appropriation of a text is to do justice to that text.

## HERMENEUTIC INTERPRETATION OF LANDSCAPE CHANGE

Ricoeur's view on the dialectic between these two ways of reading can be helpful in thinking through what a critical understanding of the meaning of a landscape might entail. Although there are of course many differences between landscapes and texts,[8] to understand our interpretational relationships to landscapes as *reading practices* emphasizes the fact that when landscapes present themselves as somehow significant, they still need interpretation. Landscapes do not "speak" to us; rather they beckon to be interpreted. Their meaning is not merely "out there" waiting to be discovered, they have to be *appropriated* by us as interpreters (Drenthen, 2011). To understand what a particular place means to me, I need to appropriate it, make it my own, and in our attempts to understand what these environments mean to us, we always take ourselves along. And yet, in doing so we also need to do justice to the landscape itself. Just as a structural analysis of a text, an ecosemiotic reading of a landscape may focus on the internal physical and semiotic relationships within that landscape. Such an analysis can focus on the other, non-human semiotic perspectives that are present in the landscape, on the relationship between different significant elements in that landscape, and on the changes of these through time. Doing so can reveal the complex context of our own understandings of the world and remind us that our human perspective on the world is not the only one.

Yet, we are ultimately thrown into our own symbolic understandings of the world, our human (all-too-human?) perspective. A hermeneutic interpretation of the landscape, on the other hand, will set out to examine how we interpreters can "appropriate" the landscape, can interpret and understand the landscape as a place that one can inhabit. Hermeneutics assumes that humans are meaning-seeking beings, and the world they inhabit is a reflection thereof. We live in a world that is always already interpreted, that is: the meanings and interpretations of our world are no secondary addition to an otherwise "objective" reality, but rather *form the very fabric* of the kind of world that matters to us. A *critical* hermeneutic understanding of a landscape, however, aims to find a deeper understanding of oneself-in-place that incorporates a structural landscape analysis.

The world that appears to us as meaningful is a world that we managed to appropriate, make our own. But every now and then an environment presents us with issues that breach our world view and that need to be acknowledged in our interpretations of the world.[9]

We interpret the world by "appropriating" it through our pre-understandings—bringing the phenomenon "home," as it were—and yet, in an attempt to do justice to our experience of meaning, we also need to distance ourselves from our pre-understandings and expose ourselves to what the "text" of the world itself has to say to us. Because of this dialogical relation between self and world, each hermeneutic understanding of the world also implies a form of self-understanding: how we interpret the landscape reveals as much about the world as it does about ourselves.

And therefore, as soon as radical landscape change forces us to revise our understanding of a landscape, our identity is at stake as well.

O'Neill, Holland, and Light point out that "people make sense of their lives by placing themselves in a larger normative context, and that this is one of the reasons why *environments* matter to people too: because environments embody just such contexts for self-understanding" (2008, 162–164). But given the complexity of the landscape, and the dialectic between reader and text, the landscape can give rise to different normative interpretations of the meaning of a landscape, which might play a role in many environmental conflicts. Environmental conflicts are not merely conflicts between different interpretations of an environment, but often also conflicts between different notions of personal identity and sense of place (Drenthen 2009a, 2013; Deliège and Drenthen, 2014). That is why different landscape readings can easily give rise to deep and seemingly irresolvable conflicts about the landscape, especially when existing landscape interpretations are challenged by rapid landscape change. This is the case even more so when we are confronted with changes that are not initiated by us, but where nonhuman nature itself spontaneously changes and forces us to redefine our place. Interpreting a place requires an active appropriation and investment of self, but also a willingness to let the text tell me something about myself that I did not know before. We must therefore understand how places have always already contributed to who we are, learn to understand ourselves *through the landscape that we find ourselves in* (Drenthen 2011), and then move on to produce more adequate interpretations of the meaning of the land to enable more adequate practices.

A critical landscape hermeneutic will stress that each narrative interpretation of the landscape has to do justice to the text itself, and thus include a relational analysis of the landscape text. When landscapes change, they force us to reconsider the old ideas we have about these landscapes. If the landscape changes and relationships between different species change, one could argue that the structure of a text changes, the text itself changes, which forces us to re-read the landscape text and acknowledge the structural changes that have taken place. This way, an ecosemiotics analysis of changing landscapes can provide a criterion for ethical debates on landscape conflicts.

Environmental hermeneutic seeks to better understand the meanings at stake in these conflicts of interpretation, and is engaged in finding ways to further the dialogue. Finding a new relationship to a text does not merely involve acknowledgment of the structural composition of the text, it also involves a change in the way that the reader is involved in the reading of the text. A critical environmental hermeneutics will therefore have to explicate the interpretational base of our being-in-the-world by articulating those pre-existing meanings and interpretations that already play a role in how we act and think, and in doing so force us to have a second look at them. Some of our previous interpretations of the land may prove to be inadequate or outdated once we properly reflect upon them.

A hermeneutical environmental *ethics* will ask in what sense these old interpretations can still be considered adequate articulations of how the world we find ourselves in beckons to be understood, or whether we should seek new articulations. Rearticulating these meanings can be laborious, but plays a critical part.

The task of a hermeneutical environmental ethics, then, is to articulate and make explicit those interpretations and meanings that are already at work in our everyday practices, to bring them to light and make them explicit, and to confront existing meanings and interpretations with other, less obvious interpretations. Doing so will increase our sensitivity for the many different meanings that can be at stake in our dealings with a particular place.

Conversely, changes in the relation between humans and nonhuman species also urge us to drastic rethinking of existing ideas. A case in point is the return of wolves, who are recolonizing thoroughly humanized lands. The conflicts between those who want to welcome the return of resurging wild animals and those who feel they should be stopped are not just about wolves, but also about different notions of environmental self.

In the following section, I will give a brief impression of how a hermeneutic of wolf resurgence would proceed, and how it would incorporate elements of a structural analysis.

## A HERMENEUTIC ANALYSIS OF WOLF RESURGENCE IN THE NETHERLANDS

The resurgence of the wolf in Western Europe is one of the most spectacular examples of spontaneous rewilding. The wolf was exterminated in most parts of western Europe in the nineteenth century. Populations only existed in more remote areas in Eastern and Southern Europe. But in the last decade, wolf populations are again spreading across Europe. Italian wolves have moved north to the French, Swiss, and Austrian Alps; Eastern European wolves have established themselves in former East Germany, and are moving westward.

In early 2015, the first wolf entered the Netherlands. It crossed the border from Germany and roamed the northern parts of the Netherlands for four days, walking for 200 kilometers, mostly through meadows and small woods, but also next to roads and highways.

The return of the wolf immediately made headline news. Initially it was met with a mixture of fascination, excitement, and anxiety; however, when the animal passed through commercial zones and occasionally through towns, many had a growing sense of unease. Clearly, the animal did not show the kind of shyness and fear for humans that the expert had predicted. Was this really a wild wolf, or might it be a wolf-dog hybrid? And might an animal with no fear of humans pose a threat? Policy makers started making contingency plans—tagging the animal, scaring it away, releasing it in less populated areas—but before these could be put into practice, the wolf disappeared again. After four days in the Netherlands, the wolf went back to Germany, and disappeared from sight.

Interestingly, the debate about the appearance of the wolf did not only concern existing ideas about wolves, and about what would be an appropriate relationship between humans and non-human nature. It also specifically addressed the relation between the Dutch and their landscape.[10] It turned out that the different views on whether or not people should welcome the wolf and should learn to share living space with them were deeply intertwined with different notions of the nature of the Dutch landscape as well as with different notions of environmental identity. Moreover, the wolf's arrival appeared to challenge existing notions both of the specific cultural nature of the Dutch landscape and of what it means to live in a cultural landscape.

In the following section I will discuss at least four different interpretations of the meaning of wolf resurgence as they surfaced in the debate: the wolf as an intruder, as an innocent victim and friend attempting to return to where it belongs, the wolf as a normal animal and an object of management, and the resurging wolf as a symbol for the return of self-willed nature.[11] Some of these wolf images, so I will argue, are dependent on specific understandings of the relation between humans and nonhuman nature, or culture and wild nature if you will, views that are no longer adequate in light of the recent developments. As a result of changes in the relational role of wildlife in the hybrid landscapes of the Anthropocene, existing wolf images are called into question and as a result existing identities are challenged too.

### *Wolf Haters*

One of the most prominent responses to the arrival of the wolf was disbelief, followed by anxiety and hostility. Many argued that the Netherlands is much too small and too densely populated for wolves to settle there. Some even made the stronger claim that a proper wolf would never *want to* come to a place like the Netherlands, therefore there must be something wrong with it. The image of the wolf as a "wild" animal and intruder that does not belong to present-day cultural landscapes is persistent (Skogen et al. 2008; Tønnessen 2010), despite convincing evidence that wolves are not confined to wild places such as old growth forests, but are perfectly able to flourish in cultural landscapes, and have done so for many centuries, as long as there is enough prey and there are enough hiding places. This image of the wolf fits perfectly into a worldview in which the separation of wild and culture is of crucial importance. In this view, cultural landscapes derive their meaning and value from their (human) history of cultivation, from having been appropriated from (a-historical) nature. The Dutch seem to take pride from the idea that their landscape is human-made. "God created the world, Netherland was made by the Dutch," so a Dutch saying goes. To cultivate a landscape is to "humanize" it, to establish meaning, and bestow sense upon the world. The implication of such a view is that cultural landscapes and wild lands are separate realms of reality.

In this view, there is no problem as long as wolves remain in the outside world of wilderness. But as soon as they appear in *our* world, they are seen as intruders. In other words, a structural analysis of wolf behavior challenges the existing self-understanding of wolf haters, and calls for revisions of deeply seated ideas about the relation between self and wildness.

What is the most disturbing thing about wolves for those who fear its arrival appears to be that the wolves themselves, through their behavior, seem to undermine the very distinction between wild and cultivated land that is so important to the defenders of cultural landscapes. By challenging the very difference between wilderness and culture that is so important to the identity of heritage protectionists, wolves present a threat to the kind of environmental identity that relies on a neat separation of these domains. The hostility toward wolves is a response to the threat of a particular normative worldview.[12]

### *Wolf Lovers*

Not all people opposed the wolf's return; many feel that wolves deserve our respect and that we should seek to coexist with them. But some go one step further and claim to have a deep emotional connection to them. In their eyes, the return of the wolf—this beautiful, intelligent, and highly social animal—is a blessing for a country where nature has almost disappeared. These wolf lovers typically talk of wolves as the victims of a hostile human culture that seeks to subdue nature. In contrast to a morally corrupt society, wolves represent authenticity, pureness, honesty, grace, and innocence: "I'd rather have a wolf as a friend"; "a wolf is honest, people lie and deceive"; "humans are the real beasts." What is striking, though, is that many self-declared wolf lovers compare wolves to their own domestic dog, and thereby systematically embezzle the basic fact that wolves are predators. Unlike domestic dogs, wild wolves need to stalk their prey, use their social skills and intelligence to hunt and kill.

The problem with this perspective is not just that it is naive, but rather that it conceives of wolves as isolated individuals, not as functioning members of an ecological community. This perspective ignores any ecological as well as ecosemiotic understanding of the wolf's place in the landscapes they inhabit. This lacking recognition of the structural role of wolves might also be the reason why initial public support for wolves can quickly evaporate and turn against the wolves as soon as conflicts appear. As long as we live separate lives, we might try to ignore their presence, fence them off, or claim to love them. By failing to recognize wolves as "emplaced" parts of an ecosystem, as inhabitants of the same landscapes that we live in, this perspective fails to see the ethical challenge in wanting to coexist with wolves.

If we do not recognize wolves as real animals living in the ecological, semiotic, and social context of our landscapes we are not doing justice to the text.

### *Wolves as Manageable Objects*

Both wolf haters and wolf lovers seem to have difficulty acknowledging the wolf in all its relevant features as part of a living landscape. For this reason, many feel that the most important thing that can help further the debate is to separate fact from fiction, and to fight prejudices and preconceptions about the wolf that are not based on facts. The assumption of this "wolf management" approach is that if we see wolves objectively, as "normal" animals, and seek a sober and balanced, rational and realistic understanding, it should be possible for a modern rich country such as the Netherlands to change its policies in such a way that humans can coexist with wolves. Such an approach makes ample use of existing ecological as well as ecosemiotic knowledge about wolves. Wolf managers rightfully assume that it is possible to correct false ideas about the wolf. However, by stressing the "normality" of the wolf, they also implicitly seem to assume that one can abstract from interpretative understandings of wolves. Such a normalization strategy for wolf perception fails to recognize the fact that perceptions of wolves are situated within a broader context of cultural ideas and normative worldviews regarding the relation between humans and non-humans. Wolves have been invested with meanings that cannot be changed at will. The attempt to manage wolf perceptions rationally and detached from emotion can be criticized for being insensitive to the many meanings wolves have—be it as evil or charismatic animals, independent and autonomous or unruly and scary. Wolf management suggests that wolves can and should be turned into "manageable objects." Could it be that the wolf management approach implicitly shares the assumption of both wolf haters and wolf lovers: that unruly wildness—as that which resists human orderings—does not have a role to play in the Dutch landscape? Underneath the managers' approach to conflicts between wolves and humans is a hidden assumption that the integration of wolves into our society does not require a radical change of the deep-felt beliefs about the human-nature relationship.

*Wolves as Symbols of the Unruly Wild*

In response to the reflex of control, some people stress that the resurgence of wolves confronts us with the need to update our ideas about our place in the natural world and urge us to put in perspective how we take human power over nature for granted. The successful resurgence of wildlife in Europe can be attributed to a shift in the dominant attitude of Europeans toward nature from hostility to tolerance. Swedish biologist Guillaume Chapron argues that "the European model shows that people and predators can coexist in the same landscapes." "I do not mean that it is a peaceful, loving coexistence; there are always problems. But if there is a political will, it is possible to share the landscape with larger predators" (cited in Conniff 2014).

Resurging wolves confront us with our desire for control, not only control over nature, but also control over our own nature. Yet, the wolf's wildness is disquieting and uneasy to some, and fascinating to others (Drenthen 2009b). Resurging wolves force us to rethink what it means to live in a living wild landscape. If we want to find an appropriate *habitus* that allows us to live together with wolves, this will require some degree of management and control. But respecting nature's autonomy also implies a willingness to live with wild creatures, not just when they are charismatic and cute, but also when they are a nuisance. Without practicing tolerance—the virtue of enduring those things that are difficult to endure—wildlife management will inevitably incarcerate wildness. Wolves force us to recognize that in our desire for control, we lose sight of the unruly in nature, the unruly that we fear and that fascinates us. They confront us with our limitations and finitude, and put at stake the image we have of ourselves, but at the same time also remind us of our deep and profound fascination for the vitality of nature.

The uneasy truth of the resurging wolves is that we have forgotten what it means to live in a world that remains to be wild. The possible return of wolves in landscapes where they were thought to have gone extinct forever, challenges existing notions about ourselves. We have to relearn *who we are* in a world that is still—to a large degree—uncontrollable and wild.

The real hermeneutic challenge of wolf resurgence is that we humans have to learn to deal with unruly animals, because we do share a world. All traditional attempts to clearly distinguish the human from the non-human were based on a violent separation between the human and the non-human world. If we really want to live together with the wolf, we need to recognize that in ecological terms we are just one species among other species, and also that there are differences that cannot be bridged.

We cannot love wild wolves the same way as we love our pets. If we as a modern society want to live together with wild wolves and share the same environment, we will have to learn to deal with the unruliness which is intrinsic to wild animals, and become more tolerant to the possible threats and discomforts that the presence of these animals can bring along. We also have to develop a form of interspecies communication in which we teach the wolves where to live and where to keep away from. We have to respect the animals, but resist our romantic longing for a harmonious coexistence without tensions. If we acknowledge that we want to live together with wolves, then we also need to acknowledge the difference between humans and wolves, and keep distance from one another.

## CONCLUSION: THE ROLE OF SEMIOTICS IN LANDSCAPE HERMENEUTICS

For Ricoeur, the world of the text is fundamentally different from the "real world" that the speaker can point toward and that both speaker and listener live in. A written text, in contrast, opens a world of the text that exists only insofar as it is interpreted by a reader. Similarly, landscape can bring to mind entire worlds: long-gone worlds of traditional farming, where humans and the land lived together in mutual dependence, or worlds in which humans had to fight against nature to carve a place for themselves in a hostile environment. These worlds only come into existence by our active interpretative acts—the readers and interpreters of this "text"—and yet these meanings are not freely invented, but result from a serious attempt to understand the meaning of the landscape as a story of what it means to live in this land.

But in a sense, the world of the landscape text is also very "real" in a literal sense. Forrest Clingerman (2004) has noted that, in this respect, the landscape is a very special kind of text, because the world that is brought forth by the landscape-text *is* the real world (in a very specific sense at least). One could say that a structural analysis of the landscape focuses the attention on these "real" features that the reader can "point out," and to the "reality" of the world of the landscape-text, in which the reader finds himself, without abandoning the need to interpret.

Similarly, a semiotic understanding of the resurgence of wolves can call our attention to how we relate to other species, and how these species relate to us. Such an analysis can bring into play the "objectivity" of the non-human beings with which we share a place.

Living together with wolves obviously requires that we take note of their way of understanding the world, so that we can take their needs and behaviors into account, and that we are prepared to engage in forms of interspecies communication. If we want to live together with wolves in a mixed community, we have to learn from and about wolves, and wolves must be able to learn from us how they can live together with us. An ecosemiotic analysis can help see the possibilities and limitations of the interspecies communication, help us understand how our actions not only have meanings for ourselves but also have significance for the non-human organisms around us, and make us aware about how our actions have changed the world of wolves. The merit of such a semiotic approach is that it shows what the field of significance in which we humans find ourselves looks like. The ecosemiotic work of Morten Tønnessen (2011), for instance, clearly shows how the semiotic relationships between humans, wolves, and sheep have shifted and changed. A critical hermeneutics must take account of those views.

Yet, in the end, the "impartial" perspective of the semiotic analysis has to be integrated with an overall hermeneutic, partially human perspective on the meaning of wolf resurgence, in which all the objective features are put into the interpretative context of a narrative and comes to *mean something*. It is through narratives that we can connect our fate with that of wolves in the landscapes we inhabit. Environmental hermeneutics is engaged in the moral conversation about what the current situation means to us humans, and what this situation requires from us. And that conversation is a matter of humans amongst each other. An ecosemiotics understanding of the role of wolves in the landscape can make clear *what* the challenges are, and *how* they can be met, but a hermeneutic of wolves is needed to find a reason *why* we should respond to them.

## NOTES

1. The scientific community had been talking about climate change much earlier, obviously.
2. Ellis & Ramankutty (2008) show how most of the biomes of the Earth have been impacted by human activities.
3. Another great example is Morten Tønnessen's ecosemiotic analysis of Norwegian wolf management (Tønnessen 2010, Tønnessen 2011).
4. Aristotle already made a similar distinction between the kind of knowledge that humans share with animals that refers to pleasure and pain, and other meanings directly related to actions and sensory experiences, and the typical human knowledge or logos, that is interested in ethical meanings and assessments regarding "the just and the unjust," "good and evil." "Whereas mere voice is but an indication of pleasure or pain, and is therefore found in other animals (for their nature attains to the perception of pleasure and pain and the intimation of them to one another, and no further), the power of speech is intended to set forth the expedient and inexpedient, and therefore likewise the just and the unjust. And it is a characteristic of man that he alone has any sense of good and evil, of just and unjust, and the like, and the association of living beings who have this sense makes a family and a state." (Aristotle, *Politics* Book 1, chapter 2).
5. Williams 1985, 119–119. Also see Van Tongeren and Snellen 2014.
6. Contrary to romantic hermeneuticists, Ricoeur holds that the meaning of a text does not coincide with the intention of the author, if only because some texts accommodate much richer readings than the author intended.

7. But this does *not* imply, of course, that the meanings we encounter are made or constructed by us. The world outside exists, and throws its questions at us. See Drenthen 1999.

8. One obvious difference is that a text is fixed, whereas landscapes are constantly changing. The analogy should not blind us to the dynamics of landscapes.

9. As John Berger puts it: "What we habitually see confirms us. Yet it can happen, suddenly, unexpectedly, and most frequently in the half-light of glimpses, that we catch sight of another visible order which intersects with ours and has nothing to do with it." (Berger 2001: 10)

10. In the following I focus on contemporary imagery of wolves in the Netherlands as it played out in public debates in media in 2012–2015. Of course, older and more general cultural meanings attached to wolves are still at play as well. For a concise review of the cultural and historical meanings attached to wolves ("cultural semiotics") in Europe, see Tønnessen 2010: 56-64. Also see Drenthen et al 2015.

11. See Drenthen 2014 for a more elaborate version of this analysis.

12. The situation in the Netherlands seems rather different from that in Norway. Tønnessen notes that in Norway, it is the "idea of the changing landscapes as symbolic of rural troubles and the loss of traditional livelihoods that is fuelling, reinforcing, the wolf's negative symbolicity and the apparently never-ending conflict on wolf management." (Tønnessen 2011:73). In contrast, in the Netherlands, most of the opposition against wolves in the Netherlands is unrelated to any perceived threat to livelihoods, but rather stems from a vague sense of fear for the life of one's children and pets. This fear is not based on any actual experiences but fueled by the traditional worldview where wild humans have to fence out nature in order for humans to flourish.

# EIGHT

# Behaving like an Animal?

## *Some Implications of the Philosophical Debate on the Animality in Man*

*Carlo Brentari*

The hierarchical situation of the biotic community in contemporary times, the Anthropocene, pivots on the principle of the superiority of man over non-human animals; on this principle is based the claim that only humans are worthy of moral consideration and legal protection. At a common sense level, this idea is frequently taken as fact; from a philosophical and anthropological perspective, it shows up in a plurality of aspects and is differentiated in cognitive, moral, juridical, and political representations. In all its forms, however, this idea is impressively widespread and is rooted in the way most Western people perceive themselves. If we take a closer look, we see that it rests on ontological notions which are mostly implicit and unexpressed in common sense, and are more explicit and clear in philosophy, anthropology, and juridical and political thought. The superiority of man, in other terms, is seen as depending on the presence of an element, structure, or instance that is ontologically different from all other entities in the world, which grounds and justifies any other form of superiority.

However, the situation is more complex. In many cases—and, again, from common sense to philosophy—the principle of the superiority of man coexists with the apparently opposite conviction of the existence of a deep core of animality in every human being. Consequently, the hierarchical relationship between humans and non-human animals duplicates itself within man, where a lower, animal layer and an upper, properly human one are supposed to be found. In addition, if the animality of man is often conceived as starting from the model of non-human animals, the opposite can also happen; our views on non-human animals can be built starting from convictions and prejudices about the alleged "animal nature" of man. The latter often include the assumption of his innate aggressiveness, selfishness, and destructiveness and, in particular historical contexts, can be used as explications for genocides and mass wars (as we will see in the last section of the present chapter).

In its philosophical and most self-conscious version, the conception of the double nature of man takes the form of a problematic coexistence of *two separate functional systems*: sensitive and rational soul, instinct and reason, reflex behavior and intentional action. Philosophers, such as Aristotle, Aquinas, and Descartes, were deeply persuaded that, in particular situations (physical or mental illness, moments of strong passion, somnambulism, etc.), the lower, "animal" functional system can take over the control of human action, normally ruled by reason, and make humans "behave like beasts." The present chapter analyzes two moments in the history of this conception: its birth, mostly due to the changes made by Aquinas to the psychology and anthropology of Aristotle, and its full affirmation by Descartes, where the lower functional system is thought of in a decidedly mechanist way. In the last part of the chapter, a review of modern authors (Konrad Lorenz, Frans de Waal) is proposed that shows that this model is a long-lasting one and that, in modern times, it has taken the form of the so-called "theory of the cultural veneer." In the conclusion, theoretical disadvantages affecting the model of the two functional systems will be highlighted, in particular the risk that the fear of losing our rational control can shed a negative light not only upon many spontaneous behavioral modalities of man, but also upon animals themselves.

## ARISTOTLE: TWO FUNCTIONAL SYSTEMS AND AN INTERMEDIATE FACULTY

The first part of our inquiry will be devoted to tracking the idea of the two functional systems in selected thinkers of the Western tradition. Our goal is not to demonstrate that this idea is central in their philosophical conception, nor that they all share the same vision of animal and man, but rather to show the historical pathway of a long-lasting model, whose influence is mostly implicit, but nonetheless extremely stark.

In Aristotle's *On the Soul*, the high biological and anthropological relevance of the soul consists of its being the functional principle of the body: "if then one is to find a definition which will apply to every soul, it will be the first actuality of a natural body possessed of organs" (1975a, 69). The soul is also defined as the entelechy of the living body, the organic unit of a body and its essential form; soul is what organizes and structures matter so as to make a specific organism.

If the soul is the organizing principle of the organism, to understand its nature in more detail, we must ask which faculties it confers to the various kinds of living beings. It is at this point of the investigation that the hypothesis emerges of a plurality of souls, or at least of functions of the soul. According to Aristotle, three souls or faculties of the soul exist in the realm of the living: the vegetative, the sensitive, and the rational. The presence of even a single of these faculties is enough to justify the qualification of an entity as a living being; the vegetative soul is to be found in plants, animals, and men; the sensitive only in animals, and men, while the rational one is exclusively in men and constitutes the specific difference of human beings. This plurality is made necessary by the empirical observation that living things perform a variety of essential functions irreducible to one another; and, if the vital activities of plants are limited to nutrition and reproduction, those of animals include local movement, the perception of objects through different sensory channels, and (as we shall see) the power of imagination. In addition to these faculties, humans are also endowed with reason, thinking, and moral skills.

Being immaterial, the three souls cannot be observed directly, but only from the vital functions with which they provide living beings; analogously, in Aristotle's view, even the vital functions cannot be understood abstractly, but only in their essential relationship to the objects that enable them to be implemented: "logically the exercise of their function comes before the faculties themselves. And if [. . .], even before these functions, the objects corresponding to them [should be examined], then for the same reason one must first of all determine the facts about those objects, for example, about food or the object of perception or thought" (1975a, 85). This three-stage explication—from the souls (or the faculties of the soul) to their functions and activities, and from those to the correlative objects—is also the key of the Aristotelian ecology. In his view, the complexity of a living being depends on the presence of one, two, or all souls, and every soul allows for the relationship between the organism and a specific typology of objects: the vegetative soul with food, the sensitive with sense-perceivable entities (and, as we shall see, to their images), the rational soul with intelligible entities. In humans, who can establish a relationship with all three objectual typologies, the complexity of the living reaches its climax; anyway (as seen in plants), every soul is complete in itself and is a fully working functional system. In Aristotle's view, the idea is already to be found that a living entity builds an inseparable unity with the environmental objects, and that every intraorganic "structure" or "force" within the organism reveals its meaning only in relation to external entities. This is not to say that Aristotle had fully developed the notion of a species-specific environment; this concept will emerge only in the twentieth century, when it will be realized that every single species has access, in his perception and action, to an *Umwelt* that is different from that of other species (Uexküll 2010; see also Brentari 2015; Canguilhem 2008). However, it is important to recognize that the distinction of three souls proposed by Aristotle is founded on the relationship to different objects, so as to make the souls *functional systems* that allow for the cognitive and operative insertion of the organisms in different ontological domains.

Let us now briefly discuss the three typologies of objects Aristotle sees as natural correspondents of the three souls: food, what is sense-perceptible, and what is intelligible. Because the difference between man and animals is linked to the definition of the higher souls, we cannot, here, face the matter of the Aristotelian theory about food intake, which opens a philosophical debate about assimilation that continues in modern and contemporary ontological reflection (see, for instance, Hartmann 1950; Johannsen 2012, 116–127). As far as perceptible entities are concerned, Aristotle includes among them all objects that can be touched, seen, heard, "felt" through taste and smell; in this typology, he includes even the state of quiet or movement of perceived things. As far as the intelligible objects are concerned, they are accessible only to humans and are primarily represented by forms and abstract concepts. In Aristotelian physics, form is the ontological component that supervenes on matter to constitute a concrete object and that, in a gnoseological perspective, allows for the conceptual inclusion of things in a certain category. The second, cognitive function comes about only through man, whose rational soul is able to grasp the general form of things. Abstract concepts (Aristotle gives the example of the "immeasurable" and of the diagonal) are mathematical, geometrical, and ideal notions that man forms by abstraction from multiple perceived objects or events. For man, then, to be related to the sphere of the intelligible objects means to overcome the domain of immediate perception and to open the way to new logical and cognitive possibilities: the intuition and formulation of general relationships between things, and the identification and expression of environmental traits which, by their nature, cannot be sense-perceived—for example, incommensurability, and then all negative qualities (such as the *absence* of a color), or purely relational entities (as the *border* between two things).

According to Aristotle, the sphere of the intelligible forms is not accessible to non-human animals. As a possible consequence, animals would be condemned to the passive reception of sensorial stimuli and to the reaction to what is immediately present. However, this is not the case for Aristotle. Perhaps in order to account for complex animal performances, or for the fact that often animals do not just react but spontaneously undertake searching activities (which will later be called "explorative behavior" (Lorenz 1981, 325)), Aristotle says that animals, just like men, are endowed with of the power of imagination (*phantasia*). In the terminology adopted in this chapter, what Aristotle calls imagination is an intermediate functional system that is common to non-human and human animals, and to which even human beings can resort, especially in situations of suspension of the rational soul's activity.

According to Aristotle, imagination is "the process by which we say that a *phantasma* [image] is presented to us" (1975a, 159); this process of image formation, as Schofield remarks, "includes cases which are not obviously instances of mental imagery, but seem more like examples of direct sensory experience" (1995, 250). In other words, more than high mental performance, imagination is a sort of first elaboration of sensations, which allows for the permanence of the sensory data even in the absence of the perceived object. At the same time, imagination is not the same as the act of isolating the pure form or the general essence of things; in this sense, it can be called a "non-paradigmatic sensory experience" (Schofield 1995, 263). Imagination is "different from both perceiving and thought; imagination always implies perception, and is itself implied by judgement. But clearly imagination and judgement are different modes of thought" (Aristotle 1975a, 157). On the other side, imaginative acts occur "only [. . .] in connection with what is perceptible" (1975a, 161) and, therefore, belong to the powers correlated to the sensitive soul.

Therefore, the function of imagination for living beings is not abstract thinking (which does not mean that, in humans, imagination can be preliminary or functional to thought (Aristotle 1975a, 177; Frede 1995, 290), but rather a guiding role within the environment). Imagination helps the organism in the satisfaction of basic needs: "without *phantasia* the desire would be without direction, hence even primitive animals have to have imagination at least 'indefinitely'; they have to aim at something" (Frede 1995, 290). Therefore, it is consistent that, among the primary duties of animal imagination, there is the search for food; if the sensitive soul were only passive sense-perception, and if no perceivable trace of food (smell, etc.) were available in the environment, an animal would not start to spontaneously seek food because, as Frede correctly remarks, "the cognitive power of sense-perception is narrowly limited to what is immediately perceived" (1995, 283; cf. also Johannsen 2012, 207, 211).

Finally, the products of imagination share two common properties with those of the rational soul: they allow for a greater plasticity of behavior than mere sensibility does, and they can be true or false depending on the quality of the basic perceptions ("all sensations are true, but most imaginations are false" [Aristotle 1975a, 159]). The last property is very useful for understanding the complex behavior of animals; for example, the search behavior: the power of imagination presents to the animal some images of food that, Aristotle writes, "are similar to objects perceived except that they are without matter" (1975a, 181). Such *phantasiai* arise "while sense-perception is still in operation and the object is present" (Frede 1995, 284), but they last in the sensitive soul of non-human animals even in absence of the object, as "after-images" that "may vary in complexity and duration" (Frede 1995, 285). It should be noted that this solution does not at all imply that the animal can grasp the abstract forms of the objects, nor construct abstract concepts. Moreover, in the Aristotelian view, not every animal species has imagination. If sensation and imagination were identical in actuality, then imagination would be possible for all creatures; however, this appears not to be the case. For instance, it is not true "of the ant, the bee, or the grub" (Aristotle 1975a, 159). Frede implicitly criticizes this passage by stating that "even a worm has to have a kind of notion of its aim in its search for food" (Frede 1995, 290; cf. the different evaluation of bees by Descartes).

Even man can orientate his actions based on images; in his case, however, this behavioral mode implies the renouncement of the rational control of action: "men often follow their imaginations contrary to knowledge, and in living creatures other than man there is neither thinking nor calculation, but only imagination" (Aristotle 1975a, 187). In other words, those who follow the images in the same way as non-human animals do, hand over the leadership of their conduct to a functional system that, in man, should be subject to reason. This can happen for various reasons: "because imaginations persist in us and resemble sensations, living creatures frequently act in accordance with them, some, *viz.*, the brutes, because they have no mind, and some, *viz.*, men, because the mind is temporary clouded over by emotion, or disease, or sleep" (Aristotle 1975a, 163).

In this regard, Schofield notes that "*phantasia* is a necessary condition of animal movement" and that, "as opposed to *noēsis*, [it] is the prerequisite of desire in non-rational animals and in what prompts the fevered or the weak-willed man to act" (1995, 255). Although essentially correct, the use of the term "weak-willed" highlights a problem that will draw our attention in the following sections; is the shift to the lower functional system (here, from *noēsis* to *phantasia*) due to a purely cognitive dysfunction (maybe of physiological origin), or is it due to a state of moral weakness? In the case of Aristotle, the first alternative is probably correct; as we have seen, the distinction among the three functions of the soul rests on the distinction among the corresponding environmental objects that are to be known, and (at least in *On the Soul*) the question of individual will, or moral conduct, is not significant. As we will see, the "moral turn" in the discussion on the plurality of the souls—and, consequently, on the possibility for human animals to shift to animality—will be due to Aquinas. Closer to the Aristotelian letter is the paraphrases proposed by Frede: "[Aristotle's] definition suggests that *phantasiai* are mere after-images of sense-perception, often false ones [. . .], which guide animals since they do not have reason, and human beings when they are disturbed by passion or disease, or are asleep" (1995, 280). In any case, for Aristotle, imagination makes at least some animal minds closer to the higher faculties of human beings; it is an intermediate power that bridges the gap both between the two main parts of man (sensitive and rational soul, "animality" and "humanity"), and between human and non-human animals.

Many contemporary studies in psychology, ethology, and cognitive science show an increasing interest for imagination, frequently with the aim (reminiscent of Aristotle's view) of finding common psychological, behavioral, and/or cognitive traits between human and non-human animals (Mitchell 2002; Taylor 2013a). Starting primarily from Darwin's observations on dreams (1871/1896), and in a lesser number of cases from Aristotle (1975a), these studies investigate the function of imagination in the behavioral repertoire of many social and/or explorative animal species: image formation and usage in problem solving (for instance, in shaping a tool), creativity as the capacity to see the perceptual reality in alternative ways, pretension in apes (frequently studied in comparison with pretend play in children), planning capacity, and deception (the latter as realization of the failure of a plan). In many cases, scholars usually define imagination as a faculty with two sides, separating the more elementary process of the formation of mental images out of perceptual data from higher faculties, such as planning and pretension ("imagination can occur as mental imagery, or more creatively, in perceiving something as something else" (Mitchell 2013, 1)).

Great interest is also shown toward the neural correlates of imagination (for a review, see Taylor 2013b, 5).

## THE SEPARATION OF THE TWO FUNCTIONAL SYSTEMS IN AQUINAS

In Christian theological thought, the Aristotelian conception undergoes a deep interpretative change that is derived, in its core, by a new way of conceiving the soul and its substantiality. According to Aristotle, the soul is a substance only inasmuch as it is the functional unit of body and soul, which raises the question whether or not it is separable from the body and, consequently, eternal and incorruptible. The problem is faced by Aristotle with the assumption that only a part of the intelligible soul, the so-called productive intellect (Aristotle 1975a), is separable and incorruptible. This assumption is not due to moral or eschatological requirements, but only to the fact that, in order to establish a cognitive relation with the intelligible entities, the soul must be of the same nature, that is, incorruptible, pure, and eternal.

In Christian anthropology, and in particular in Thomistic thought, this "environmentalist," object-centered approach disappears almost completely. What decides about nature and function of the soul are no longer the (natural or logical-ideal) entities with which it is in a daily exchange, but its inclusion in the moral and eschatological relationship with God. The soul becomes an entity created by God out of nothing and assigned by him to every single human being. Therefore, it becomes the subject of a personal story of sin, redemption, or damnation—which logically requires that it must survive the death of the individual body; and this is possible, in turn, only if the soul is thought of as ontologically simple, immaterial and, thus, incorruptible. Indeed, God, one of whose aspects is absolute justice, cannot punish a person whose identity changes substantially in the period subsequent to the fault. Although immaterial, the Christian soul must also be a substance in a sense that can be briefly defined as "thing-like": a separate entity created by God to be a stable and self-identical moral subject. The idea of the soul as a functional unit with the body, although present, has now a secondary importance.

Let us now see what this implies for our discussion about the alleged animals, animality, and the alleged animality of man. In the *Summa Theologica,* the status of non-human animals is not specifically addressed but can be clearly deduced from a series of passages in which Aquinas uses the difference from animals to speak about man and his metaphysical destiny. This requires a deep modification of the Aristotelian frame. First, the distinction between the sensible and rational soul loses its cognitive character and becomes a prevailing moral and eschatological issue. For example, while describing the situation of the world after the last judgment, Aquinas examines the question of the permanence or disappearance of animals in it. In this context, the philosopher states that, after the "renewal" of the judgment, animals disappear, because "dumb animals, plants, and minerals, and all mixed bodies, are corruptible [. . .] both on the part of their matter which loses its form, and on the part of their form." On the contrary, humans are corruptible only as for their matter, not as for their form: "the rational soul [. . .] will remain incorrupt after the corruption of man" (2013, q. 91).

The ontological separation of human from non-human animals is also reached in another way that is consistent with the transition from a cognitive to a moral approach. In the *Summa Theologica,* human subjects are characterized by the possession of free will and by the possibility to determine autonomously their own conduct; on the contrary, animal behavior is seen as hetero-directed and subjected to mechanical influences. The performances of animals, whose complexity led Aristotle to admit the pre-rational faculty of imagination, are explained by Aquinas through the notion of instinct (*instinctus*). This term, which is already in use in the works of Fathers of the Church, will have a long history and will be responsible for many misunderstandings about animality. As Diamond correctly remarks, in Aquinas, the term "instinct" is not used exclusively for animal behavior, but refers generically to a broad set of hetero-directed actions, from heavenly inspiration to the influences of the stars; instinct is, therefore, "a motivational term, which [. . .] usually points to an external source of motivation" (1971, 324). With few exceptions, this broad use has nowadays disappeared, and today the term refers to those inner, mostly mechanical drives toward different objects (sexual partners, prey, etc.) whose presence we assume in human and non-human animals. Even if it could seem contradictory to use a term originally referring to an external source of motivation to indicate an inner drive, it is evident that the reference to compulsion and hetero-direction has not disappeared, even from the contemporary meaning of the notion.

Nevertheless, let us now come back to the Thomist text. While discussing the possibility to practice divination by reading the flight of birds, Aquinas talks about the principles of animal behavior. After having recognized that "the movements or cries of birds [. . .] are manifestly not the cause of future events," and, therefore, that "if anything future can be known from them, it will be because the causes from which they proceed are also the causes of future occurrences," he writes that "the cause of dumb animals' actions is a certain instinct whereby they are inclined by a natural movement, for they are not masters of their actions." This instinct "is subject to the disposition of surrounding bodies, and primarily to that of the heavenly bodies. Hence nothing prevents some of their actions from being signs of the future" (2006, II-II, q. 95). The same idea of the radical hetero-direction of animal behavior is expressed in the context of a discussion about murder. In this *quaestio*, Thomas considers whether the killing of non-human living beings is or is not considered a sin, and states that it is lawful to kill them because "dumb animals and plants are devoid of the life of reason whereby to set themselves in motion; they are moved, as it were by another, by a kind of natural impulse, a sign of which is that they are naturally enslaved and accommodated to the uses of others" (2006, II-II, q. 64). The thesis of the hetero-direction is confirmed by the observations of Kenny on the precise character of the actions of animals: "Aquinas thinks that the actions of animals are voluntary, but only in a diminished sense of the word. [. . .] Voluntariness, he says, involves an internal origin for an action, and a degree of knowledge of the end. [. . .] He is not denying that [animals] aim at goals by doing things in order to achieve them; but he denies that this is a choice of means to ends" (1993, 82).

The hetero-direction of animals, paradigmatically expressed by the statement "they are moved, as it were by another," opens the problem of explaining complex animal behavior without making use of any superior cognitive faculties. The position of Aquinas is the assumption of a sort of innate natural knowledge: "birds naturally know certain things regarding future occurrences of the seasons [. . .]; the turtle, the swallow, and the stork have observed the time of their coming. Now natural knowledge is infallible and comes from God" (2006, II-II, q. 95). With this statement, another long-lasting element is added to the instinct-based model of animal behavior: the infallibility of the innate knowledge on which instinctive behavior should rely. This new element is somehow necessary because, according to Aquinas, "dumb animals have naught but a sensitive soul," and a sensitive soul "every power of which is the act of a bodily organ" (2006, II-II, q. 95). Here, the power of imagination (which, according to Aristotle, was plastic and efficient, but in no way infallible) seems to play a very minor role.

A similar mechanization of the Aristotelian conception of the animal also occurs in the case of memory. According to Aristotle, some non-human animals are endowed with the power of memory, which is part of their sensitive soul. Memory cooperates in particular with the sense of hearing and increases their intelligence: animals endowed with it "are more intelligent and capable of learning than those which cannot remember" (Aristotle 1975b, 3). The status of memory, reminiscent of that of imagination, is therefore that of a power intermediate between pure sense-perception and intellect. Let us see how Aquinas paraphrases the Aristotelian passage quoted above: "prudence is found in irrational animals: hence it is said in the beginning of Metaph. A, 1 that 'those animals which, like bees, cannot hear sounds, are prudent by instinct.' We see this plainly, in wonderful cases of sagacity manifested in the works of various animals, such as bees, spiders, and dogs" (2006, I–II, q. 13). The decisive insertion of the "prudence of instinct," which replaces imagination and memory, reinforces the mechanization of animal behavior.

Consequently, the effectiveness of animal behavior is ascribed to an external source of rationality: God. This is done through the analogy of artificial things ("clocks and all engines put together by the art of man") and natural entities: "as artificial things are in comparison to human art, so are all natural things in comparison to the Divine art. [. . .] And thus it is that in the works of irrational animals we notice certain marks of sagacity [. . .] ordained by the Supreme art" (2006, I–II, q. 13). For Aquinas, animal behavior is a hetero-directed, but efficient and far-reaching, system; the entity that organizes this system is not an autonomous subject, as the sensitive soul is exposed to a long series of factors: God's rationality, inner "instincts," and the influences of the body (and, through the body, even planets and stars).

When analyzing the transition from the Aristotelian to the Thomist conception of animality, we focused more on the changes than on the traits of continuity, pointing out that, with Aquinas, animals lose autonomy and capacity for independent action. In any case, inasmuch as it credits to non-human animals a purposeful behavior, instinct is a functional analogue of reason the same way that, in Aristotle, the combination of sense-perception and imagination was. In other words, both for Aristotle and for Aquinas, humans are endowed with two different functional systems, of which the lower one (respectively, sense-perception plus imagination, and instinct) is normally subordinated to the higher one. Moreover, for both thinkers, there remains the possibility that, in particular conditions, humans can fall back into the lower organization; according to Aquinas, this happens, for instance, in the absence of positive laws and institutional rules: "under the state of the Law of nature man was moved by inward instinct and without any outward law" (2006, III, q. 60); and anyway, states Aquinas, "our will to be happy does not appertain to free-will, but to natural instinct" (2006, I, p. 19).

## THE ANIMALITY IN MAN: A LONG-LASTING THEORETICAL MODEL

With Aquinas, the sensitive soul becomes a passive, hetero-directed domain of the instincts; in this way, animals are "mechanized" and deprived of that flexible and rich possibility of action and experience that was made possible by the Aristotelian powers of imagination and memory. At the same time, a view of man is proposed wherein instinct and reason are doomed to a problematic coexistence. In this model, what we have called the lower functional system does not disappear, nor is it essentially modified by the higher one; on the contrary, it remains untouched under the higher system and can reactivate when the leading function of reason ceases. In those cases, self-consciousness, behavioral autonomy, and even moral responsibility are substituted by an instinctive, "animal" obedience to external or internal stimuli.

This duplicity of levels will provide a very long-lasting schema to modern and contemporary authors; after the disciplinary separation of science from theology and philosophy, it will be adopted even in biology and ethology, where it sometimes plays the role of a pre-comprehensive frame, of an occasionally dangerous basic assumption that may influence the interpretation of empirical data. We cannot examine every version of the two-level model as proposed by later authors, nor mention every critique of it; instead, we will focus on two final problems linked to the concept of instinct and to the assumption of the animality in man.

## DESCARTES: MECHANIZATION OF ANIMALS AND AUTOMATISMS IN MAN

The first of the two problems is that, in its Thomistic use, the concept of instinct was not able to explain the complexity of animal behavior in a satisfactory way. Moreover, as we have seen, Aquinas used the term "instinct" in a generic and broad sense and was not directly interested in studying animal behavior (his purpose was rather to emphasize the moral and ontological superiority of man). Consequently, the true extent of the risk of mechanization associated with the term of instinct emerges only later in philosophers who are sincerely interested in explaining animal behavior. One of the clearest examples is Descartes, who uses mechanism as a comprehensive theoretical frame for understanding nature and, as far as animals are concerned, brings the idea of instinct to extreme consequences by integrating it into the conception of the animal as an automaton. What is new, as compared to the Thomistic model, is that the functioning of the animal-machine allows for greater behavioral precision, and sometimes even for the occasional superiority of animals to man.

In the letter to the Marquis of Newcastle on November 23, 1646, Descartes observes that even if animals "can do many things better than we do," this demonstrates only that "they act naturally and mechanically, like a clock which tells the time better than our judgement does" (1970, 207). In this passage, Descartes refers to some behaviors that strike for their apparent intelligence and purposefulness: the migration of swallows, the organization of a bee hive, the discipline of cranes in flight, the order "of apes in fighting, if it is true that they keep discipline," and finally, the impulse of apes "to bury their dead," that "is no stranger than that of dogs and cats who scratch the earth for the purpose of burying their excrement" (Ibid.). All these cases of teleonomic behavior are traced back to the model of the clock; their complexity, and even their superiority to human performance, is due to mechanization and the absence of judgment.

In Descartes, the long-lasting influence of the model of the two functional systems is not limited to the analysis of animals. Under particular conditions, human behavior (which is normally intentional and self-conscious) can fall back under the control of the lower system. Indeed, in the letter mentioned above, Descartes states that animals "only imitate or surpass us in those of our actions which are not guided by our thoughts. It often happens that we walk or eat without thinking at all about what we are doing; and similarly, without using our reason, we reject things which are harmful for us, and parry the blows aimed at us." According to Descartes, the particular conditions under which humans act without using the higher functional system can occur in somnambulism ("those who walk in their sleep sometimes swim across streams in which they would drown if they were awake"), in reflex actions ("even if we expressly willed not to put our hands in front of our head when we fall, we could not prevent ourselves"), and in conditions of strong passion ("as for the movements of our passions, even though in us they are accompanied with thought because we have the faculty of thinking, it is none the less very clear that they do not depend on thought, because they often occur in spite of us" (Ibid., 206)).

This is a critical point in our inquiry. The modern version of the model of the two functional systems leads to a situation where, on the one side, animals can act more effectively than humans when they follow their instincts and, on the other side, when their reason or judgment is suspended, humans can behave like animals or machines (which is the same, in the Cartesian view), and sometimes even with extreme effectiveness. This approach has two dangerous epistemological consequences. First, one could believe that the analysis of the biological phenomena in which humans act following the lower functional system allows us a glimpse into the inner organization of animal life; second, the entire animal experience could appear similar to the condition of a somnambulist, or to a continuous subjection to passions, without the mitigating influence of thought.

Here we cannot follow in all its consequences the connection between instinct and reflex action, introduced by Descartes and shared by many others, such as Herbert Spencer, Heinrich Ziegler, Ivan Pavlov, Jacques Loeb, and many others (Brigandt 2005). It should be enough to say that, if the Cartesian analogy with the clock and the implicit reference to the divine Clockmaker slowly disappears from the scientific debate, the concept of reflex (with its many related notions: stimulus, automatic reaction, conditioned and unconditioned behavioral modes, etc.) will have an enormous fortune. A systematic criticism of this concept will be undertaken only in the twentieth century, for instance, in Merleau-Ponty's *The Structure of Behavior*. After having recognized that the concept of reflex has its roots in the Cartesian mechanism (in particular, in Descartes 1985), Merleau-Ponty explicitly states that "far from being a faithful description of behavior, the theory of conditioned reflexes [. . .] carries over into organic activity the modes of cleavage which are appropriate to a universe of things" (1963, 56).

## INSTINCTS BOILING UNDER A VENEER OF CIVILIZATION

The second theoretical problem linked to the model of the two functional systems (instinct and reason, "animality" and "humanity") emerges with the connection of the higher system with the notion of civilization. Relying on the primatologist de Waal, this version of the model can be synthetically defined as that of the "veneer of civilization" (2005, 19). Its birth is inextricably linked to the particular situation of the second post-war period:

> With its gas chambers, mass executions, and willful destruction, World War II was human behavior at its worst. [. . .] Comparisons with animals were ubiquitous. Animals lack inhibitions, the argument went. They lack culture, so it must have been something animal-like, something in our genetic map that had burst through the veneer of civilization and pushed human decency aside. This 'veneer theory,' as I call it, became a dominant theme in the post-war discussion. Deep down, we humans are violent and amoral. (Ibid.)

According to the veneer theory, not only is there animality in man, but under particular conditions, the lower, animal-like functional system can reactivate and allow for a sort of blind and cruel efficiency of human action, even in the absence of rational guidance (much like Descartes' somnambulists). Only if the animality of man is seen as a separate but efficient system, can it be indicated as responsible for the bursts of brutality that we encounter in human ethology and history. This assumption of an original, instinctive natural aggressiveness that is "inhibited" by culture, and in particular by modern civilization, can be found in many ethologists, primatologists, psychoanalysts, and psychologists of the twentieth century: in the meta-psychological works of Freud (1958; 1961), the ethology of Lorenz (2002), the crowd psychology of Allport (1924) and Le Bon (according to which the mostly unconscious dynamics of the crowd would remove the cultural superstructure interiorized by individuals, and let out the instincts in their purest form; 1896, 13), and in many others.

All activities of mediation and defusing conflicts are not natural; they are part of an artificial veneer of culture, which brakes and hides the original magma of instincts and aggressive tendencies—including the tendency to relate to neighboring populations or to ethnic minorities as if they were beings of another species, a phenomenon that goes under the name of pseudo-speciation (Erikson 1966; Lorenz 2002, 76–80). This model establishes a conflictual relationship between instinct and culture, based on the opposition between the "original" and the "derivative," and considers institutions, rules of coexistence, and even laws as fragile reaction-formations that cannot change human nature but only keep it at bay until the next outbreak.

In the second post-war period, the great diffusion of this explicative model has led to numerous attempts to understand what might be, as if in the pure state, the animal nature that humans hide beneath the friable mask of culture. Some of the most important results have come from comparative research on primates; as de Waal points out, in the eyes of many educated persons apes seem to illustrate with utmost clarity what the original human nature could be. In particular, the social behavior of chimpanzees seems to reflect our animal core; research on chimpanzees conducted by Goodall in the 1970s led to the discovery of cases of infanticide and even real wars between neighboring groups, with incursions and organized raids. In 1979, reports de Waal, "the posh pages of *National Geographic* reported that these apes also kill each other, sometimes consuming their victims" (2005, 24). One of the bloodiest conflicts involved "a community [that] split into a northern and southern faction, eventually becoming separate communities. Those chimpanzees had played and groomed together [. . .] and lived in harmony. But the factions became to fight nonetheless. Shocked researchers watched as former friends now drank each other's blood" (de Waal 2005, 135). The verdict was unanimous: humans and apes wear the mark of Cain as killers of their own kind; even primatology, in conclusion, seemed to reinforce the idea of the original animality of man (including the tendency toward intraspecific violence), which could be contained by means of appropriate educational and cultural strategies, but not permanently removed.

Since the 1980s, the model of the cultural veneer has been questioned by several authors. The feminist philosopher of science Haraway (1992, 2003) developed an evolutionary perspective centered on the notions of partnership, companionship, and cooperation between humans and non-human animals. In ethology, the studies of Bowlby and Eibl-Eibesfeldt (the latter a student of Lorenz) showed that attachment and empathetic and supportive behaviors were as original and widespread as aggression and competitive behavior (Bowlby 1969). In particular, Eibl-Eibesfeldt enclosed the empathetic behaviors in what he calls "synagonal strategies," which he places on a plane of equality with the agonal system of the aggressive tendencies (2007, 520). In primatology, scholarly attention to the bonobo (or pygmy chimpanzee) refuted the idea that the entire evolutionary branch of apes is characterized by a high rate of intraspecific aggressiveness. Indeed, bonobos, although territorial like chimpanzees, can avoid violence between groups using sexuality as a way of peace-keeping. When two groups of bonobos meet, the females of the two groups open up the possibility of mediation through grooming and rubbing the genitals, and in a short time, two groups of males and females mate freely. Only the adult males of the two groups remain unwelcoming to each other, but the freedom to interact with the females of the other group prevents this latent hostility from resulting in action.

The ease with which new empirical data on bonobos have led to a reversal of the relationship between "natural" and "derived" (from the priority of aggression to the model of the "good savage" who avoids violence in the most spontaneous way) shows us that this kind of model is problematic in itself. In other words, in the face of empirical data, too sharp theoretical contrapositions—original vs. derivative, human nature vs. culture, animality vs. humanity—appear severely inadequate, often because they are vitiated by a preliminary ruling pessimism concerning man. One should remember that, if instincts are seen as the contrary of morality and rational self-control, this sheds a negative light not only upon the spontaneous behavior of man, but also upon animals themselves, which are frequently seen as the "wild side" of humans, with all the moral prejudices that may result.

## THE QUALITATIVE ALTERITY OF MAN AND, THEREFORE, OF ANIMALS

Having followed, in a necessarily partial way, the philosophical debate on the animality in man, we can now suggest some concluding remarks. By analyzing the Aristotelian anthropology, we have seen that the idea of the two functional systems was developed in order to explain the cognitive abilities of sentient beings. Starting from the principle that the similar knows the similar, Aristotle suggested the existence of a different function of the soul for each type of object present in the environment of living beings. In Aristotle, although with some ambiguity, the idea that, in humans, two functional systems coexist has not a moral character, but a cognitive one; furthermore, the role given to imagination and memory makes animals closer to man by establishing an intermediate common level. At the same time, however, Aristotle introduces the possibility for man to pass from one to the other functional system, an idea that will have a long life and that, in different theoretical contexts, will constitute the logical prerequisite of the belief that man can act like an animal.

In Aquinas, we found two further steps of utmost importance. On the one hand, imagination loses its importance in animal behavior and animality comes to coincide with the mere, bodily-conceived perceptive soul; on the other hand, in order to escape the problems arising from that limitation, Aquinas introduces the concept of instinct, which should explain the complex behavior of animals. While succeeding (at least partially) in this task, the resort to instinct condemns non-human animals to hetero-direction, if not to behavioral passivity. Consequently, the assumption of an animal layer in man results in the conviction of a blind instinctual substrate, an idea that the philosopher associates with sin, absence of free will, and moral enslavement. In addition, the possibility of the shift of the functional system is considered in a double, cognitive and moral, perspective; and there is no doubt that Aquinas' attention is more focused on strengthening the rationality and morality of man than on studying animal behavior as an independent field of inquiry.

In Descartes, the clear ontological separation between *res cogitans* (the individual soul) and *res extensa* (matter, including organic matter) leads to a radical mechanization of animals. In the letter to the Marquise of Newcastle, in which Descartes answers direct questions about animal behavior, he denies animals any cognitively superior ability and sees the lower functional system as a well-planned mechanism. The hetero-direction of non-human living beings, already introduced by Aquinas, is here enhanced in two ways: on the one hand, the mechanistic physiology of Descartes leads to the consideration of animal behavior as a series of mechanical reactions to external stimuli; and on the other, the perfection of animal mechanisms is traced back to the omnipotence of God's reason. Moreover, by carrying forward the idea that man can switch from one to the other functional system, Descartes introduces a relevant analogy between normal animal behavior and human reflex behavior: they both occur in the absence of rational control and are characterized by a kind of blind effectiveness. Unlike Aquinas, in analyzing the difference between animals and humans, Descartes does not have primarily a moral purpose; his interest in physiology is authentic, and the model he proposes (mechanism) will for centuries be considered to be a valid key for interpreting organic matter.

The model of the two functional systems, which combines a clear separation between humans and animals with the occasional possibility for man to fall into the lower functional system, has revealed itself as long-lasting. In the twentieth century, its most significant variant is the theory of the veneer of civilization: instincts (especially aggression and sexuality) lead an underground existence under a civilized and moral layer; if this latter loses its leadership, they are ready to take control with the blind effectiveness that is their own. In the cultural atmosphere of the second half of the twentieth century, this model seemed to be the most suitable to explain historical-ethological phenomena such as genocide and mass war (whose effectiveness, incidentally, has probably more to do with the modern technical means than with the alleged animality of man). In any case, in certain authors (as in Lorenz), this idea is accompanied by the belief that the instinctual substrate is an indispensable part of man; other authors, like de Waal, propose instead a critical assessment of the model itself.

The philosophy of the twentieth century provides at least two possible alternatives to the dualism of the functional systems. The first is offered by German philosophical anthropology, with authors such as Scheler, Plessner, and Gehlen (Gehlen 1988; Plessner 1975; Scheler 2009). According to them, the difference between humans and non-human animals is so radical that they do not share any common behavioral traits, let alone an entire functional system. The core idea is that the qualitative alterity of human animals, although fully natural, coincides with the absence of fixed instincts and with the necessity of the cultural elaboration of environmental stimuli (see, in particular, Gehlen 1988). Consequently, in the hypothetical situation of the absence of culture, humans would not fall back into a lower functional system, nor act instinctively: they would be completely unable to act.

The second alternative is proposed by von Uexküll, particularly in *Theoretical Biology* and *A Foray into the Worlds of Animals and Humans* (1928; 2010). Uexküll not only rejects the prejudice of the two functional systems, but also adopts a horizontal consideration of all animal forms. His analysis of the species-specific functional circles (which correspond to different subjective *Umwelten*) highlights that different behavioral possibilities are present at every level of the animal sphere. In Uexküll's theory, the refutation of the traditional functional dualism does not lead to a culture-based conception of the qualitative alterity of man (as in the German philosophical anthropology), but opens the possibility of an equal, non-hierarchical consideration of all semiotic and pragmatic environments that can be found in nature.

This is not the place to go into detail about how these two alternative approaches work. The present chapter was dedicated to the starting point and key moments in the history of the model of the two functional systems in the Western philosophical tradition, and to the two main dangers that its uncritical adoption by philosophers, ethologists, and scientists in general can bring with it: the danger of considering man as a "bad" animal, to which is added an ontologically higher and, therefore, "good" instance, and that of considering non-human animals as men acting without the faculty of rational self-control. It is enough to say that (in different ways) both alternative approaches avoid these dangers, and that, in criticizing the often unexpressed ontological and axiological pre-comprehension of human observers, they open the way to a more equalitarian and empathic view of non-human animals.

# NINE

# Seeing with Dolphins

## *Reflections on the Salience of Cetaceans*

*Katharine Dow*

One of the most striking presences when I first went to North Eastern Scotland in 2005 was that of bottlenose dolphins. I was some months into my fieldwork before I saw a live one, yet still dolphins were everywhere—in advertisements for cetacean-watching boat-trips, in the small blue plastic dolphins that many people hang from the rear-view mirrors of their cars, in the craft products sold in local shops. In casual conversations, people talked about dolphins with admiration and awe. They shared tips about where to spot dolphins from the land and celebrated sightings whenever they occurred. When I nosily looked through the windows of people's houses while out walking or going to the shops I would regularly see sculptures, paintings and photographs of dolphins inside.

There is a resident population of one hundred and thirty bottlenose dolphins in the Moray Firth, the triangle of sea that juts into northeast Scotland whose vertices are marked by John O'Groats to the north, Inverness to the west, and Fraserburgh to the east. Minke whales and porpoises are spotted regularly and there are occasional sightings, and strandings, of other species. Whenever I walked along the shore or drove in and out of Spey Bay parallel to the beach, it was hard not to let my gaze drift out to sea—perhaps I was missing a breaching dolphin this very moment! In Spey Bay, people keep one eye on the sea, an ear out for news of sightings. Bottlenose dolphins, a relatively rare, wild species which hold a great many positive cultural meanings, are thought of as something that makes this area special. Visitors come and spend all day waiting, watching, wondering, and hoping. Looking out to sea is a highly contemplative experience. Like stargazing, it invites reflection, and I sometimes used to wonder if other people used watching for dolphins as an excuse simply to gaze out into the blue-grey yonder and to let their minds wander, as I often did.

In this chapter, I will focus my attention on a specific group of people in this area who work and volunteer in the wildlife center in Spey Bay, a tiny coastal village on the southern Moray Firth coast, where I carried out ethnographic fieldwork between 2005 and 2007. The wildlife center is run by an international cetacean conservation charity and represents the locus of concerns about whales and dolphins, as well as anxieties about the environment more generally. It is open, free of charge, to the public, who are invited to watch for dolphins, whales, porpoises, seals, and osprey from the beach[1] and they provide information through exhibits and talks about the local wildlife and the plight of cetaceans to visitors and organized school groups.

The humans who work in the wildlife center all have slightly different ideas about and varying commitments to environmentalist campaigning and animal rights, yet they share the belief that the environment is subject to anthropogenic climate change and a concern about the future of the natural world, epitomized by the varying and in many cases acute threats to cetaceans. So, while they did not use the term "Anthropocene" themselves at the time, they do agree that humans are having a potentially permanent and devastating effect on the planet—including in particular the marine environment. The Save the Whale campaign was one of the first modern manifestations of the Green movement and cetaceans retain a prominent role in environmentalism[2]. In the UK, attitudes toward cetaceans are metonymic of attitudes to the environment over the last century—cetaceans have gone from being a resource to be exploited, to impressive creatures in need of protection from and by humans. This is as true in northeast Scotland, once a hub of commercial whaling (and since the 1970s, the center of the North Sea oil and gas industry) but now somewhere to which people travel to see cetaceans in the wild, as anywhere.

Here, I will explore what it means for humans to have close relationships with wild marine animals and how human-dolphin relations playfully trouble species boundaries and complicate notions of (human) nature, wildness, and anthropomorphism. As the title of the chapter suggests, most of the closeness that people in Spey Bay achieve with dolphins is visually mediated and I will discuss the reasons why people in Spey Bay promote seeing as the most ethical form of engagement with dolphins later in the chapter. Of course, many people long for other forms of closeness with dolphins and auditory communication has played a vital role in the history of humans' relationships with cetaceans, for example in scientific experiments attempting to teach dolphins to "speak"[3] and in the huge positive impact that the popular dissemination of audio recordings of humpback whale song had on the campaign to end whaling (Rothenberg 2014). In this chapter, I prioritize vision, as that is how people in Spey Bay get close to dolphins, and because, in this case, seeing dolphins helps people relate to them. Through the case of the people who live and work in Spey Bay, I want to suggest that seeing (with) dolphins is not only about appreciating charismatic megafauna, magical escapism, or getting closer to nature, but also about asking ourselves tough questions about what kind of place dolphins, and humans, will occupy in the future. I will ask what dolphins mean to humans, how those meanings are formulated, and how people formulate themselves, and dolphins, through this meaning-making activity.

## ANTHROPOLOGY, ANIMALS, AND THE ANTHROPOCENE

In her foundational analysis of the racial and gender politics of primatology, Donna Haraway stated that "We polish an animal mirror to look for ourselves" (1978, 37). The "we" in this case was North Americans in the latter half of the twentieth century, but her point can, I think, be extended to the United Kingdom and perhaps other parts of Europe and North America. As Haraway pointed out, "people like to look at animals, even to learn from them about human beings and human society" (1978, 37). So, while looking at animals is in one way an attempt at seeing beyond the borders of the human, of confronting otherness, it can be just as solipsistic as any other form of anthropology. Anthropology by definition errs toward the human, so although animals might be "good to think" (Lévi-Strauss 1964), the thinking we are doing is, more often than not, about ourselves. I would like to use Haraway's more-than-human reflective tool to consider, in this chapter, the polishing work that goes both into looking for animals and seeing ourselves reflected back in their mirrors.

Molly Mullin (1999; see also 2002) invokes Haraway's specular trope in her review of human-animal relations in socio-cultural anthropology, which looks at the "windows and mirrors" that anthropologists have employed to analyze these encounters. Mullin traces a shift, as we reached the current millennium, from looking at human-animal relations to better understand the binary oppositions of nature/culture, female/male, wild/domestic and so on toward a view of non-human animals and their relations with humans as "dynamic" (1999, 2). Looking at non-human animals through human eyes allows for a more nuanced take on human understandings of nature and the natural world and of the realities of inter-species identification. As Anna Tsing (1995, 136-7) has put it, "We know what is natural by comparing ourselves to animals. . . . Yet our views of nature are not a simple reflection of our valued standards and ideals: our observations of non-humans present continual challenges to our cultural agendas that require new inflections and transpositions of our cultural 'sense'."

Having once focused on the legacies of colonialism and modernism, anthropologists, Mullin wrote in 1999, have become more likely to engage with moral and political questions about how best to treat animals. She predicted that we would see increasing numbers of ethnographic studies that considered the question of how humans engage with non-human animals in an era of sustained attention to the state of the natural world. One demonstration of the accuracy of this prediction is recent work on the complicated local and international politics and economics of cloning animals (Franklin 2007), including endangered species (Heatherington 2008; Friese 2013). Despite the apparent novelty of cloning techniques, responses to them tend to reignite familiar questions about local, national, and species identities, the intimate embrace of domestication and capital, the ethics of creating life in labs, visions of the future (and mythologies of the past), and the workings of authoritative knowledge about nature and culture.

In many cases, debates about such biotechnological interventions into animal life have stimulated reflections on who "we" are, as individuals, nations, and communities, as Sarah Franklin (2007) shows in her analysis of the multiple meanings of sheep—cloned and otherwise—in Britain's colonial history and contemporary identity (see also Cassidy 2002 on horseracing, class, and kinship in England and Thompson 2002 on the multiple meanings of elephants). Yet, as Tracey Heatherington (2008, 10) notes, "the objectification of endangered species as genetic resources" through practices like seed banking or the creation of Frozen Zoos which store the genetic materials of endangered animals, "tends to decontextualize them further from locally embedded and culturally meaningful patterns of human-animal relations." As all this suggests, the politics and ethics of animal-human relations are more complicated than questions of animal rights or environmentalist ethics, because non-human animals are implicated in the myriad ways in which we make sense of the world, from gender to nationalism to science.

Just over a decade after Mullin's essay was published, Eben Kirksey and Stefan Helmreich (2010, 545) declared that "A new genre of writing and mode of research has arrived on the anthropological stage: multispecies ethnography." This is a form of ethnography that considers the ways in which humans become commingled with microfauna, flora, and uncharismatic critters: insects, microbes, and fungi are now receiving the attention long granted to agricultural animals, pets, and charismatic megafauna. Though, just because the anthropological gaze has been extended to less familiar species, this does not mean that the overarching questions or aims of this ethnography are necessarily that different. As Kirksey and

Helmreich (2010, 545) put it, "multispecies ethnography centers on how a multitude of organisms' livelihoods shape and are shaped by political, economic, and cultural forces." Yet, they also argue that multispecies ethnographies are moving away from the "windows and mirrors" approach toward *living with*—rather than just thinking with—animals (2010, 552; see also Haraway 2008).

Contemporary ruminations about animals within socio-cultural anthropology reflect an atmosphere of heightened attention to the ethical and the natural. Anthropologists look into the world and see people increasingly concerned about the plight of various animals and the effects of humans on the natural world, as encapsulated in the concept of the Anthropocene.[4] But, as Kirksey and Helmreich note, there is some discomfort about the Anthropocene concept amongst multispecies ethnographers, as it can seem a "bit too anthropocentric" (2010, 558). To avoid reinstating humans at the center of analysis, they suggest that we use the insights of multispecies ethnography to rethink and undo received categories—and hierarchies—of nature and culture (2010, 562–563).

As both Sarah Franklin and Donna Haraway point out, when we look at humans' relationships with non-human animals, we need to think about the ethics of situation. That is, we should be "exploring not only what are the questions to be asked" about the human-animal border, "but how are they situated, and the politics of their location" (Franklin 2011, para. 1). Whichever species they contain, ethnographies write culture and all ethnographers must grapple with the responsibilities of representation, which may be all the more acute when ethnographic subjects cannot speak back (Kirksey and Helmreich 2010, 563). Both the rise in multispecies ethnography and the concept of the Anthropocene lead to questions about how humans are thought to think about other species and the natural world more generally, and the effects of representing those thoughts.

In the next section, I will outline how the people who work in the wildlife center in Spey Bay think about dolphins, focusing particularly on shifting and slippery notions of wildness, nature, and species difference. In the subsequent section, I will extend this description by considering the different ways in which humans seek closeness with dolphins whilst retaining an awareness of their wildness and the ethical dilemmas that that presents. I will then explore how dolphins help frame and garner support for environmentalist ethics and act as a model for human behavior toward the environment. Finally, I will consider what ethnographic studies of human-animal interactions bring to the concept of the Anthropocene and vice versa by asking what model of human nature is reproduced by the Anthropocene.

## WHAT DO WE SEE IN THE DOLPHIN MIRROR?

### *Dolphins Recognize Themselves in Mirrors . . .*

Given the non-human species under discussion in this chapter, it is tempting to point out the anthropocentrism of visual metaphors—after all, dolphins mostly "see" through sound waves. But, they are one of the few non-human animals that respond to mirrors. That is, like some great apes, they appear to recognize themselves when humans invite them to look at a piece of glass with a reflective backing (Sickler et al. 2006; Marino 2013). This titbit of scientific knowledge is just one of the many tantalizing morsels that marine biology offers up to the cetacean-curious looking for evidence for their conviction that dolphins are special. It also hints at the ways in which dolphins forcefully confront humans with both their sameness and their difference.

As most readers will probably be well aware, in Europe and North America, dolphins have particular and largely positive cultural meanings. The main characteristics that Euro-American humans associate with dolphins are high intelligence, sociability, peacefulness, and even altruism (Sickler et al. 2006; Fraser et al. 2006; Servais 2005; Whitehead and Rendell 2015). Popular ideas focus on dolphins' intelligence and their capacity for communication, metonymically represented in their large brains and the sense that many humans get that dolphins are constantly smiling. These characteristics suggest a similarity between humans and dolphins, which is pushed further in the belief of some humans that dolphins are a superior species (Fraser et al. 2006), while some scientists now talk about dolphins and whales as having their own culture (Whitehead and Rendell 2015). A review of popular media depictions of dolphins in the United States identified four main themes:

1. Dolphin as peer to humans, of equal intelligence or at least capable of communicating with humans or helping humans;
2. Dolphin as representative of peace, unconditional love, or an idealized freedom in harmony with the natural order;
3. Dolphin as naïve or innocent, in which they are subordinate and vulnerable; and
4. Dolphin as superior to humans, associated with a higher power or intelligence. (Fraser et al. 2006, 327)

While particular people may find these themes more or less convincing, they are nonetheless easily identifiable cultural tropes that suggest much about the thick meanings of dolphins.

In Spey Bay, contemporary cultural meanings of whales and dolphins are largely in line with the ideas discussed above for Americans, but these meanings have certainly shifted over the centuries, as the fact that this area was once involved in commercial whaling attests. Tourism is an increasingly important part of the Scottish economy and the annual financial value of whale- and dolphin-watching tourism in the northeastern region alone is over £10 million (Davies et al. 2010). At the Spey Bay wildlife center, at least, these are not only foreign or English tourists—many visitors are from other parts of Scotland and local people would take visiting friends and relatives out on cetacean-watching tours. Survey data shows that there is generally a favorable attitude toward cetaceans and their conservation in Scotland: 80 percent of Scots surveyed supported the idea of introducing a specific law to protect cetaceans and 40 percent agreed that if a politician proposed such a law they would see him or her in a more favorable light (Scott and Parsons 2005; Howard and Parsons 2006).

I had initially assumed that the friendly association between people and cetaceans in northeast Scotland was a relatively recent phenomenon that had emerged alongside the decline in North Sea fishing and the mainstreaming of concerns about the environment. Certainly, these factors are important to contemporary ideas about cetaceans and especially to their ethical and cultural value. However, this interest does seem to have historical precedents. Marine biologist E.C.M. Parsons (2012) has argued that the number of carvings of "Pictish beasts," which appear to be cetaceans, and probably dolphins, on standing stones in northeast Scotland indicates that cetaceans have been culturally significant in Scotland since the Iron Age. Parsons suggests that Saint Columba's encounter with a strange creature, which later became thought of as the Loch Ness Monster, may in fact have been a whale in the Moray Firth. He also says that the unicorn that features on Scotland's Royal Crest is likely to be inspired by sightings of narwhals which could have ranged as far as the waters around Scotland in the chilly climate of the Middle Ages and Renaissance (Parsons 2012). Whatever the historical veracity of these claims, the efforts some contemporary researchers have gone to in order to find links between humans and cetaceans in Scotland does at the very least reinforce the deep significance that these animals have today.

The positive cultural meanings of cetaceans can be helpful to people campaigning for their conservation and environmentalism more generally, providing an attractive fleshy exemplification for their efforts. In Spey Bay, these associations are bolstered by local pride in bottlenose dolphins. Many times during fieldwork, I observed wildlife center staff telling visitors about the "family groups" and "strong social bonds" between dolphins, often painting a picture of dolphins living in piscine extended families. These kinds of meanings were particularly important in encouraging visitors to the wildlife center to sign up to the charity's main fundraising scheme, Adopt a Dolphin. In this scheme, which makes the kinship between humans and dolphins all too apparent, supporters make monthly donations to the charity and in return they receive regular magazines and updates about "their" dolphin and the charity's conservation work as well as an initial "gift" of a welcome pack with an adoption certificate, a dolphin cuddly toy, badge, and mug, all branded with the charity's logo.[5]

While staff in the wildlife center recognized the ethical and political value of dolphins' popular appeal, it also made them personally uncomfortable. Privately, they would make fun of the more "hippie" depictions of dolphins. Once, a flyer from a self-described "dolphin channel" who claimed she could talk with dolphins made its way to Spey Bay and the staff were both appalled and amused. Another time, I was standing on the beach at low tide with Nina, a member of staff, watching some dolphins leap in the firth against a pink sunset. Referencing some of the more twee depictions of dolphins available in the cultural repertoire, Nina remarked dryly that all we needed to complete the scene was a rainbow and some wind chimes.

Like Nina, most of the paid staff in the wildlife center and many of the volunteers have a background in the natural sciences and they share the assumption that whatever they think, and teach others, about dolphins should not stray too far from scientific knowledge. But, when some scientists use the terminology of dolphin "societies" and "culture" when talking about these animals (see, for example, Marino 2013; Whitehead and Rendell 2015), associations between the species become slippery and mobile. The animal mirror needs little polishing when we use the same language to talk about dolphins and humans.

One of the characteristics popularly attributed to dolphins by humans is a capacity for emotion (Sickler et al. 2006). This works both ways, as dolphins also arouse emotions in humans. Every time I observed someone watching dolphins at Spey Bay, whether a member of the wildlife center staff, a visitor to the center, or a local resident, the emotional valence of the experience was palpable and typically described by them as "awesome," "amazing," and "incredible." Despite my own cynicism about the cutesiness of many representations of dolphins when I arrived in the field, when I started to see dolphins myself, I could not deny feeling a strange sense of happy awe that never really went away with each sighting, perhaps partly because sightings of dolphins are fortuitous—the reward for being, and seeing, in the right place at the right time. Dolphins, often depicted as benign, playful, and peaceful, exemplify the category of charismatic megafauna (Barney et al. 2005). This charisma is important in securing support for the cause of their conservation and one way in which this is achieved is through eliciting empathy in humans. At Spey Bay, this is intensified by the emotional experience of watching them in the wild.

## CLOSER TO NATURE

### *Mirrors Do Not Faithfully Reproduce Our Sense of Reality, Especially under Water*

Numerous accounts of humans swimming with dolphins for leisure and/or therapy attest to the enjoyment and fulfilment that humans can experience from being close to, or touching, dolphins (Servais 2005), but such pursuits are discouraged and condemned by animal rights activists and some conservationists (Marino 2013). This includes the charity that runs the wildlife center, which argues that captivity takes cetaceans out of their natural environments and prevents them from living normal and healthy lives. For examples, captive cetaceans are forced to live in small tanks when under normal circumstances they could range up to one hundred miles a day. Dolphins often suffer stress in captive environments, which weakens their immune systems and can exacerbate aggression between individuals kept in close proximity. The means by which they are caught also often lead to the deaths and injuries of other dolphins that get caught in the path of the hunts and the youngest individuals are targeted for capture, disrupting the pod's reproductive cycle.

At Spey Bay, where it would be unwise to go swimming given the sea currents and water temperature, staff in the wildlife center encourage visual, rather than haptic, closeness to dolphins. When asked about where to see dolphins, as they regularly are, staff will advise visitors first to look from land—either at Spey Bay or from Chanonry Point, a spit of land on the other side of the inner Moray Firth—and secondly to go out on a reputable wildlife-watching tour run by operators who will not harass or distress the wildlife, that is, from which they can see but not touch dolphins.

When visitors come to the wildlife center, staff members commonly strike up conversation by enquiring if the visitors have seen whales or dolphins before, to which the answer is, sometimes, "yes, at SeaWorld."[6] Staff told me that this response presented them with a moral dilemma, because they are wholeheartedly against keeping cetaceans in captivity, yet they were wary of embarrassing or alienating visitors by criticizing their past activities. This sympathy was hardest to maintain, though, when visitors came to the wildlife center expecting to see dolphins in captivity there, which was a fairly regular occurrence. One staff member said to me in private what she would like to be able to say to such visitors: "We don't keep dolphins in tanks like SeaWorld! You have to come and watch and wait, you can't just expect to get everything you want whenever you want it!" The misperception of some visitors that they might have dolphins in captivity at Spey Bay is exasperating for the wildlife center staff, as it is a fundamental misunderstanding of what they stand for and it demonstrates that there is still a lack of awareness about the problems of keeping cetaceans in captivity amongst members of the general public.

For many humans, the more they know about the specific qualities of dolphins, the greater their compulsion to act responsibly toward them. Lori Marino, a behavioral neurologist and advocate for better treatment of dolphins by humans, has described how her research with dolphins, which includes the mirror recognition experiments mentioned earlier, led to a devastating awareness of humans' routine mistreatment of dolphins and her own responsibility to speak out against it. She writes:

> There is abundant scientific evidence for complex intelligence, self-awareness, and emotional complexity in dolphins. Dolphins possess large complex brains second in relative size only to those of modern humans. They have demonstrated prodigious cognitive abilities in such areas as language understanding, abstract thinking, and problem solving. They recognize themselves in mirrors, understand the relationship between their own body and that of others, and can reflect on their own thoughts, showing that their sense of self is not unlike our own. Moreover, their complex sociality and cultural traditions are well documented from ongoing field studies. These findings provide strong support for recognizing their status as individuals with basic rights comparable to those of humans. Yet, dolphins (and whales) continue to be treated as non-sentient objects, commodities, and resources. Egregious examples of this are the many dolphin and whale slaughters that occur around the world and the exploitation of dolphins in entertainment parks, research facilities, and the 'dolphin therapy' industry. (Marino 2013, 95)

The staff at the wildlife center in Spey Bay believe both in the impressive attributes of dolphins and in protecting them from harm. They also share with other local people a sense that dolphins are an important part of the area's identity and value. The close presence of dolphins in the sea nearby is part of what makes this a place in which they can have a good life (see also Tsing 2012, 142, on the delight of making places familiar through mushroom foraging). Yet, in private conversations they were keen to demonstrate their awareness of the fact that, while there may be similarities between humans and cetaceans, they *are* different species, not least because cetaceans are wild animals. Quite a few times, I heard staff members talking about the fact that pods of dolphins "play" by throwing and catching baby porpoises amongst themselves. Dolphins do not usually eat porpoises, so this is not hunting or feeding behavior, but is seen instead as a sort of "blood sport." They did not usually mention this in front of visitors, but on one occasion when I was volunteering with the wildlife center manager Sophie, a visitor to the center, brought it up while he was chatting with us. Sophie told him that the charity that runs the wildlife center does not deny this behavior and said, "people tend to see dolphins as these happy, friendly creatures, but they are really wild animals, and we need to remember that." The man agreed and concluded, "a lot of the time, they behave better than humans, anyway."

Dolphins are mammals that live in the water; their very nature confounds everyday assumptions. Dolphins swim in and out of species categories, just as they come up out of the water to breathe and go back into it to hunt, feed, and sleep. In the thinking of people in Spey Bay, dolphins deserve to be protected because they are both like, and not like, humans. Yet, all this sympathy toward dolphins happens from a distance, across an elemental divide that says much about the complicated, ambivalent desire to be closer to nature.

Humans attribute a complex range of characteristics to dolphins, which complicate notions of the wild and domestic, nature and culture. Reflecting on the various claims that are made on their behalf shows that sometimes the rationale for the ethical treatment of dolphins rests on their human-like qualities, sometimes on their more-than-humanness, and sometimes on their otherness. Thinking about human objections to dolphin captivity shows the muddiness of our anthropomorphism and the fluidity of our thinking about these animals (see also Chrulew 2011; Knight 2006 on parallel issues with zoos). Dolphins are seen as having "rights" for a range of reasons, from their big brains to their capacity for sociality and emotionality to an assumption about what the conditions for a good life might be for any creature. Some humans think that dolphins should be free just because they are a different species, to which wildness comes naturally. Either way, the idea of rights is a very (Euro-American) human one, as is the assumption that "freedom" and "happiness" are the goals of a good life. Is "the wild" dolphins' home or their habitat? Is captivity for dolphins harmful because it is unnatural, punitive, or damaging to their health? Can our ideas about the ethics of dolphin captivity help us gain a new perspective on human incarceration? Or should humans respect dolphins' autonomy by not interfering in their lives? These everyday dilemmas faced by people working in cetacean conservation unfold the profound questions that inform how people think, and live, with non-human animals.

## DOLPHIN ETHICS

### *Mirrors Can Withstand Extremes of Temperature, Rising Seas, and Climatic Change Better than Humans or Dolphins*

Debates about captivity are just one form of ethical engagement that humans make with dolphins. As noted, cetaceans are iconic of the environmental movement and particularly the movement to conserve endangered species. Many species of cetacean are endangered and the Yangtze River dolphin is thought to have become extinct in the last decade. Bottlenose dolphins are not, in fact, endangered, and seem to have a relatively stable global population (IUCN 2014). They are, however, a protected species under the EU's Habitats Directive and Appendix II of the Convention on the International Trade in Endangered Species of Wildlife Flora and Fauna (CITES), as they represent "species that are not necessarily now threatened with extinction but that may become so unless trade is closely controlled" (CITES n.d.). Whilst they may not be currently endangered, bottlenose dolphins in Spey Bay are compelling mnemonics of people's fears about the irrevocable damage which human activities can wreak on the planet. As the different risks of endangerment for different cetacean species indicate, endangerment has different timescales—for some, annihilation is imminent, while for others it remains a vague threat. For environmentalists, the salient point is that all these risks are serious and humans are implicated in each of them, in terms of both cause and effect.

Though critical of past and present activities, environmentalism is a future-oriented movement. One of the problems for people campaigning for better awareness of environmental problems is where to focus their efforts and advocacy—which particular part of the future should we concentrate on saving or improving? In knowing where to pitch their efforts, environmentalists and conservation campaigners must walk a thin line between demonstrating the seriousness of their cause and inspiring doomsday apathy. During my fieldwork, I helped wildlife center staff at a number of events. I was often struck by how concerned about the state and future of the environment the adults and children I met at these events were. At one event in Aberdeen, I helped supervise children making posters about threats to cetaceans and their habitats, which many of them took to enthusiastically. But, when I explained what the children were doing to a group of passing visitors, one mother with a child of about six years old wrinkled her nose and said, gesturing to her daughter, "I'm not sure I want her to know about things like that, I don't want to fill her innocent ears with things like that."

The campaigning efforts of the staff in the wildlife center at Spey Bay are focused on cetacean conservation, but many of them are equally, if not more, concerned about climate change and wider environmental threats. Dolphins and whales provide a useful hook with which to encourage people to think about the environment, and what they can do to care for it better. The case of cetacean conservation also makes the interdependence of species and the environments clear. Many of the dangers to dolphins are anthropogenic. Most obviously, these are the dangers of being caught and kept in captivity, slaughtered in drive hunts, injured by boats, or drowned or caught in fishing nets. But they are also threatened by the ways in which human activities like polluting waterways, overfishing, and emitting carbon affect their ability to survive in the wild and contribute to climate change. And, as the wildlife center staff is all too well aware, dangers to dolphins are also dangers to humans. The slippage in people's thinking about the species boundaries around dolphins and humans becomes very useful in pointing out the deep ramifications of environmental change.

In Spey Bay, dolphins as they are now exemplify a good life, both through the example of their own sociality and special qualities, but also in that their presence makes the humans who live near them feel closer to nature. The threats that dolphins face in the future remind humans of the environment's vulnerability to human actions and of our responsibilities to the natural world. In other words, dolphins are not only "good to think" with, but also to act with. Despite the fact that one lives in the sea and one on land, dolphin and human lives are intertwined, and interdependent. People in Spey Bay know that dolphin habitats should be conserved because they, and humans, are dependent on retaining the wild spaces in which they live.

For the people who work in the wildlife center in Spey Bay, dolphins represent goodness—both in their natural characteristics and because they embody the reason for ensuring a good future in which the natural world is conserved and protected from anthropogenic damage, which is good for humans as well as dolphins. For the wildlife center staff, dolphins represent a range of ethical values that help guide their own ethical self-fashioning and their encouragement of ethical dispositions in the people who visit the center. Dolphins' intelligence and sociability suggest a model for how humans should approach the world and cultivate their intraspecies and interspecies relations. Dolphins' vulnerability to human activities suggests better and worse ways for humans to behave by showing the consequences of certain actions, but also the value of wider ethical values like responsibility, care, and compassion for others.

Lori Marino (2013, 103–104) argues that humans tend to extend our moral concerns only to those we see as members of our "in-group." She links this with movements to end discrimination on the basis of race and gender amongst humans, pointing to the importance of the recognition of similarity and difference in ethical and political arguments. As this suggests, cetaceans become most morally salient when their similarities leap to human attention. Marino says, "cetaceans—probably more than any other animal—represent extremes of similarities and differences that challenge our ability to recognize them as moral equals. That is, probably more than any other animal, cetaceans are the most vulnerable and the most bewildering" (2013, 104). She concludes that the ways in which humans respond to the challenge of moral inclusivity for other species will be a way of demonstrating the cognitive and emotional superiority that we routinely attribute to our own species. So, if we follow her line of argument, one way of being human is to recognize our kinship with other species. When humans learn to see cetaceans, we will also discover our humanity.

## SALIENT REFLECTIONS

Human attempts to treat dolphins more ethically in the context of a changing climate raise the question of what, or *who*, they need protecting from. The people who work in Spey Bay are neither misanthropic nor militant, but they are critical of human activities that egregiously contribute to the degradation of the environment. They are sympathetic to the everyday difficulties individuals face in trying to lead their lives in ways that are considerate of environmental needs, since these are the same dilemmas they face in their own lives. Nonetheless, they see dolphins as in need of human help because of the ways in which human activities impinge upon their ability to live in the wild. This suggests a divided view of humans, or human nature that is trying to reconcile optimism about our capacity for care with a perceived tendency toward short-termism and self-interest.

In a recent lecture at the University of Cambridge, Anna Tsing reflected on the figure of the *anthropos* in the Anthropocene through the trope of "Earth stalked by Man." In her lecture she asked what model of "Man," or human nature, the Anthropocene invokes. The Anthropocene is problematic for anthropologists—and feminists—because it rests on an assumption of a generalizable *anthropos* (see also Lorimer 2015). Making claims for the whole species of Homo sapiens threatens to undo the meaningful specificities that we encounter through studying animals, human or otherwise. Man is "both limited and everywhere" (Tsing 2015), yet we should be careful about reproducing and rewriting Man into the concept of the Anthropocene. The greater attention to humans' effects on the environment and other species that the concept of the Anthropocene represents is both welcome and overdue. We must deal with climate change and environmental destruction in a way that is critical of attempts to master the planet for the short-term benefit of our own species and we must be wary of apparent solutions to the Anthropocene and our changing climate that do not radically alter the ideas and behaviors that have caused it. But, we must do so whilst retaining a sensitive appreciation of diversity in all species, including humans. Misanthropy is not the solution to climate change.

Donna Haraway (2015) has recently suggested that we think of this time as a boundary event, rather than an epoch. As she reminds us, we must remain critical of the workings of capitalism, the logic of extractivism and the aim of mastery over nature whilst retaining a sensitive appreciation of diversity in all species, including humans. While I am hesitant to fully embrace the second half of her rallying cry of "Make Kin, Not Babies!" I do agree with her, and Anna Tsing, that our task—by which I mean *we* who are concerned about the environment, *we* who believe that the two main problems facing the world today are inequality and climate change, and that they are closely entwined—*our* task is to seek out, reinstate, and create those places of refuge that will help prevent and salve some of the losses we face if we continue the way we, as a planet, are going. Reflecting on dolphins, on what they mean to us, how they relate to us and the dangers that they face is just one compelling way of making these potential losses salient.

We polish the animal mirror to see non-human animals as they really are, to get to know another being across the species divide. Yet, this polishing work often makes them appear more like ourselves. Perhaps it is just that the most recognizable creatures are also the most visible. Looking into the dolphin mirror and at the reflections that gaze back shows that even charismatic megafauna can challenge the ways in which we see, or think we see, ourselves and our relationships with nature and the natural world. They also show us how such conceptions change: in little more than a century, dolphins have gone from being a resource for humans to exploit, to victims of human greed, to potential guides to save the world from impending climatic disaster. In Spey Bay in the early twenty-first century, dolphins represent what humans want themselves, and their world, to be. Thinking about dolphins is a way of expressing what they value and what they fear. Above all, they provide a salient, visual reminder that humans' lives are intertwined with other species' and that exploring relations of different and similar, antipathetic and companionate, forms of life through ethnography opens up worlds of possibilities.

## NOTES

1. They have also recently added a live video camera with a screen inside the wildlife center so that people can watch for dolphins from inside, which is considerate in a place with an unpredictable climate.
2. See, for example, the recently released documentary on the early years of Greenpeace, *How to Change the World* (Jerry Rothwell, 2015).
3. One infamous example of this was the 1960s NASA-funded project to teach bottlenose dolphins English at a Caribbean facility under the leadership of neuroscientist John C. Lilly and anthropologist Gregory Bateson. The story is told in the documentary, *The Girl who Talked to Dolphins* (Christopher Riley, 2014).
4. I cannot recall anyone in Spey Bay ever using the term "Anthropocene" during my fieldwork, which took place a few years before the concept hit the public consciousness.
5. The initial gift has changed to a cotton shopping bag and car sticker for adults and a set of stickers and a badge for children since I finished my fieldwork.
6. SeaWorld is a Florida-based chain of marine parks that display dolphins, including bottlenose dolphins and orcas, in the United States. It is not illegal to keep cetaceans in captivity in the United Kingdom, but the animal rights regulations are stricter and there is stronger public opposition to keeping cetaceans in captivity, which led to the last British dolphinaria closing in the 1990s. See Davis 2006 for more on SeaWorld and its marketing.

*IV*

# New Beginnings

# TEN

# Out of the Metazoic?

## *Animals as a Transitional Form in Planetary Evolution*

*Bronislaw Szerszynski*

In *The Natural Contract* (1995) Michel Serres argued that "man" is becoming a physical variable, weighing down on the Earth in masses, volumes, and plates which intervene directly in the operation of the Earth system. Serres suggested that this sheer size and power of humanity meant that it is starting to balance the rest of the Earth—a balancing of one species with all others which is of course an overbalancing. More than half of all accessible fresh water is now used by mankind; more nitrogen is now fixed synthetically and applied as fertilizers in agriculture than is fixed naturally in terrestrial ecosystems (Crutzen and Stoermer 2000); and human-caused erosion and denudation of the continents exceeds "natural" sediment production by an order of magnitude (Zalasiewicz et al. 2008). Around half of the Earth's land surface has been transformed by cultivation and grazing, the majority of the biosphere has been transformed into intensively used "anthromes," where natural systems are embedded in and shaped by human systems, and human beings now appropriate around 35–40 percent of the planet's production of phytomass (new plant tissues) through photosynthesis (Ellis 2011; Millennium Ecosystem Assessment 2005; Smil 2011). Returning to Serres' original image of the sheer weight of humanity, Vaclav Smil has given us the chilling calculation that humans now make up 30 percent of the mass of land vertebrates, and their domestic animals a further 67 percent, leaving wild animals constituting only 3 percent of terrestrial vertebrate zoomass (Smil 2011).

It was statistics such as these that prompted scientists Paul Crutzen and Eugene Stoermer (2000) to suggest that humans have become the determining geological force on the Earth, increasingly bending all of its systems to their own purposes and thus pushing the Earth into a new epoch of geological time: the "Anthropocene." Indeed, Charles Langmuir and Wally Broeker (2012) elevated the stakes even higher, by wondering whether we are entering not the Anthropo*cene* but the Anthropo*zoic*. The 4.5-billion-year history of the Earth is conventionally divided into four great aeons—the Hadean, the Archaean, the Proterozoic, and the Phanerozoic. Each aeon is further subdivided into eras, the eras into periods, and the periods into epochs, such as the Holocene and the proposed Anthropocene. Langmuir and Broeker's alternative name for the new unit of geological time would elevate the status of the current transformation of the Earth from that of a change of epoch *within* the Phanerozoic to a change of aeon *from* the Phanerozoic.[1] They suggest that the rise of intelligent life, a global civilization, and the technological ability to sense and direct the dynamic evolution of Earth systems has produced a potential for planetary change "almost as great as that caused by [the] origin of life or the rise of oxygen" (Langmuir and Broecker 2012, 645). They thus suggest that the transformations described in my opening paragraph might signify the planet moving into a whole new logic of development—in which case, calling this event a mere change of geological epoch would be seriously misleading.

In this chapter I take seriously Langmuir and Broeker's idea that the Earth may be entering a fifth geological aeon, but suggest that, in order to understand the nature of this shift, the category of being we first need to interrogate as a geological force is not so much *human being* but *animal being*. Only by an investigation into the "event" of animal being in the Earth, I argue, will we be able to understand the character of the aeon that may be passing; but also only through such an investigation can we start to understand exactly how the animal—even in its absence—might also shape the aeon to come. This is not to dismiss the importance of more immediate, normative inquiries into what is happening when one animal so dominates the systems of the Earth that it seems to turn all other non-human animals into relics of a bygone aeon, to be permitted to live only on its own terms. However, viewed in the context of deep, geological time, the animal becomes a figure not just of lament, but one that enables us to ask a whole other, larger set of questions about the current transformation of the Earth.

For the majority of the philosophical treatment of animals focuses on what the animal means *for the human*. The language of animal rights might try to go beyond the utility of animals to humans, but it still focuses on how our own, human, behavior ought to be modified by consideration of what animals are and what we thereby owe to them. Other philosophers have asked different, more ontological questions about animals that ask what an animal *is* (Agamben 2004; Derrida 2008). But their analyses tend still ultimately to ask about the relation between the human and the animal—even if this distinction is seen as internal to the human itself. But rather than asking what the animal means for the human, let us try to ask what the animal means *for the planet*. Undertaking this shift in the way we think about animals and the Anthropocene will not invalidate philosophical questions about the human-animal relationship—indeed, as we shall see, the question of the human *as* animal will return in a way that might force us to rethink the nature of the Anthropocene, and to conceive of the possible futures of the Earth in new and challenging ways.

## THE ANIMAL

Let me first define what I mean by an animal. In *De Anima*, Aristotle (1961) reasoned that whereas plants have a vegetative soul (in that they are capable of nourishment and reproduction), non-human animals have an animal soul (which also involves the powers of being able to sense and move), and humans have a rational soul (giving them not just those lower powers but also those of speech and abstract thought). Today the generally accepted classification of living things—which of course now includes the microorganisms that were unknown in Aristotle's day—divides them into two "domains"—the prokaryotes and the eukaryotes—further divided into five "kingdoms."[2] All life is based on self-reproducing, metabolizing cells enclosed by a membrane, but whereas the oldest two kingdoms of the archaea and bacteria are *prokaryotes*, based on rather undifferentiated cells, and are generally *unicellular*, the other three more recent kingdoms of animals, plants, and fungi are *eukaryotes*—in that their cells have a nucleus and various organelles including the mitochondria and chloroplasts—and are typically *multicellular*, in that the organisms are composed of many cells that are genetic clones but have taken on particular functions within the organism.

I am choosing to refer to animals as "metazoa." Ernst Haeckel coined this term in 1874 to refer to multicellular animals, as contrasted with the "protozoa," which he defined as single-celled animals. The term "protozoa" has fallen into disuse, being merged into the kingdom of bacteria, making "metazoa" a synonym for the kingdom "animalia," the latter now more tightly defined as organisms that are multicellular and have differentiated tissues, are motile (at at least one stage in their development) and have a digestive tract. But we can also usefully draw on the way that the "meta-" prefix can signify "beyond," or self-referentiality. Taking "meta-" in this sense, metazoa are life forms constructed from other life forms—life composed from life, life beyond life. In some sense all multi-cellular life is metazoic in this sense—animals, but also plants (which Haeckel called "metaphyta") and fungi. Indeed, Margulis (1998) and other proponents of symbiogenesis argue that life is *always* compositional in this sense, from the once free-living structures that compose the eukaryotic cell, up to the members of an ecosystem. But here I will reserve the term "metazoa" for its more conventional referent of multicellular animals.

As well as the term "metazoan" for a singular animal and "metazoa" for the plural form, we can also use the term "*metazoic*" as an adjective to refer to the state of being a metazoan—but also as a noun to refer to the current aeon in Earth's development.[3] The official name of this aeon, "Phanerozoic," means "visible life," in reference to the arrival of organisms that were large enough, morphologically complex enough, and hard enough to leave clear fossils in the rock record. For the start of the Phanerozoic coincides with the Cambrian explosion, an apparently discontinuous event which saw the rapid emergence and diversification of multicellular, eukaryotic animals (and later plants and fungi), and the establishment of most if not all the phyla and body plans that exist today.[4] Referring to the Phanerozoic aeon as "the Metazoic" would thus in one sense be a shift that is no shift at all, simply specifying the form of life—the metazoa—whose remains become so visible in the rock record corresponding to that time. But in another sense it would foreground even more sharply the metazoan, the multi-cellular animal, as not just the most geologically visible form of the fourth aeon of the Earth, but the form which defines the aeon's specificity, and perhaps its lasting significance. It is certainly true that animals were not simply something that happened to emerge in, adapt to, and benefit from the distinctive conditions of the Phanerozoic—from the huge amounts of standing biomass, the complex food webs, and the shift to rapid evolution, ecological succession, and biogeographical variation. Apart from being the first of the multicellular kingdoms to emerge, and even though (on the land at least) their mass was to be swamped by that of the plants, their presence played a crucial role in developing the complex ecological nets of the current aeon. As Nicholas Butterfield puts it, "animals figure disproportionately in the maintenance of the modern Earth System, not least because they invented it" (Butterfield 2011, 87). But in the next section I will try to explain more exactly what I mean by situating the metazoa as forms in planetary time. For, by asking what animals mean "for the planet," I am not simply trying to make an ecological move, using "planet" in the common vernacular sense of referring to the integrity of the global biosphere, but to open up to a more expansive set of questions.

## PLANETARY EVOLUTION AND THE EMERGENCE OF FORM

This resituating and restating of the question of the animal could be seen as part of a wider project of establishing the "planetary humanities." While sharing many features and sensibilities with the "environmental humanities," a planetary humanities approach would also have a number of features that would make it distinctive. For example, it would be *volumetric*, concerned not just with ecological relations on the surface of the Earth but with relations within and across all the different entangled volumes of the stratified Earth; it would be temporally extended into *deep time*, situating contemporary environmental, technological, and social changes on geological and astronomical timescales; it would involve the investigation of *the planet* as a specific category of being, getting beyond discourses of the Earth's boundedness, fragility, or uniqueness to a more general account of the distinctive ontology and temporality of planets as evolving assemblages; and it would be engaged with the *interplanetary*—both in the sense of a comparative sensibility to the divergent paths that can be taken even by similar planets, and in an orientation to the multiple ways that the stories of individual planets might become intertwined.

Above all, the approach I am taking here is concerned with the distinctive topological relations and modes of existence made possible by planetary becoming, including elemental compartments, new compositions of matter and energy, new surfaces and gradients, new entities, and new kinds of relation. I approach the Earth as an assemblage which has evolved in an ongoing dialectic between the intensive (differences and gradients) and the extensive (form and structure). The Earth has evolved in complexity through a progressive unfolding of singularities, a cascade of symmetry-breaking bifurcations (DeLanda 2002, 17, 20) that took it from being an undifferentiated cloud of atoms and particles to something progressively more anatomical or geometrical. It did this—and continues to do so—through processes of *migrating* (where constituent materials move within the body of the Earth) but also *folding* (where the Earth creates new topological relations with and within itself) (DeLanda 2002, 52); through near-equilibrium processes such as sorting and sedimenting (DeLanda 1992, 142–3) and far-from-equilibrium processes of self-organization (Bak 1996).

So from a shapeless body dominated by the *intensive*—by gradients of temperature, pressure, density concentration, and so on—the Earth developed *extensive* structures, form—what Gilles Deleuze and Felix Guattari (1988) call "strata." But the emergence of this extensivity, this internal structure, also creates *new* intensivities, new gradients, allowing the generation of new subsystems, new forms of becoming, and so on. The ongoing result is a body which is topologically complex. It thus has its well-known "spheres" such as those of rock, water, air, life, and the relations and dynamics that have emerged within these spheres and between them. But it has also created new forms, with new properties and new "modes of existence" (Latour 2013; Simondon 1958), some of which in turn may create possibilities for new self-organizing processes to be established and thus further steps in emergence. The project of classifying these new entities is a complex and open-ended one, to say the least—not least because what count as significant forms and modes of existence can change dramatically when the planet moves from one paradigm (Haff 2013) to another, and because the direction of that emergence is underdetermined, in ways we only dimly understand.

A significant contribution to identifying the major events in the unfolding self-organization of the Earth was made by John Maynard Smith and Eörs Szathmáry in their work on the major transitions in biological evolution (Maynard Smith and Szathmáry 1995; Szathmáry and Maynard Smith 1995). They argued that, rather than being a smooth and continuous process governed by an enduring set of laws, there have been some highly significant moments of transition in the evolution of life on Earth—changes produced by the evolutionary process that altered the very units of evolution and the mechanisms by which evolution proceeds. While confined to biotic evolution, this problem formulation is otherwise very close to our question of what matters "for the planet" in its self-organization; Maynard-Smith and Szathmáry and those responding to their work are asking not what moves the planet up a pre-set, developmental trajectory (since we know so little about the possible developmental trajectories of planets), but simply what enables the planet to *do new things*. The major transitions that Maynard-Smith and Szathmáry proposed are as follows: from replicating molecules to populations of molecules in compartments; from unlinked replicators to chromosomes; from RNA as gene and enzyme to DNA and protein (i.e., the emergence of the genetic code); from prokaryotes to eukaryotes; from asexual clones to sexual populations; from protists (early eukaryotes) to animals, plants, and fungi (cell differentiation); from solitary individuals to colonies with non-reproductive castes; and from primate societies to language-using human societies (Maynard Smith and Szathmáry 1995; Szathmáry and Maynard Smith 1995).

Before we focus on the emergence of metazoa, it would be useful to dwell on some general features of these or other candidate "major transitions." Firstly, all transitions of this magnitude exhibit "contingent irreversibility" (Szathmáry and Maynard Smith 1995, 228) or hysteresis: they were not inevitable, but once they occurred their significance was such that a reversion to the earlier logic was unlikely or impossible. Secondly, some, at least, represent the emergence of new hierarchical layers of association, and of new "evolutionary individuals" (such as the emergence of eukaryotic cells in a merger of unrelated microorganisms, or that of plants, fungi, and animals from genetically identical cells). Thirdly, some concern a complexification of information and meaning. For Maynard-

Smith and Szathmáry, this is about increasing the "bandwidth" and "reliability" of inheritance, through the development, for example, of DNA, sex, or language. But using the concepts of biosemiotics we could say that a better way of capturing this common feature is as the increase of "semiotic freedom" (Hoffmeyer 1996). With the very origin of the genetic code, crystalline structures themselves gained a new level of semiotic freedom through achieving what Jacques Monod calls "gratuity"—"the independence, chemically speaking, between the function itself and the nature of the chemical signals controlling it" (Monod 1972, 78). Monod uses the example of the role that different proteins play in regulating the synthesis of enzymes in the lactose system, but suggests that gratuity had the effect of "giving molecular evolution a practically limitless field for exploration and experiment" (Monod 1972, 79). But in the course of subsequent major transitions, not only does semiosis develop new levels of freedom from its material substrate, and new levels of meaning; it also starts to have a greater effect on the course of evolution itself. As Jesper Hoffmeyer surmises, "the anatomical aspect of evolution may have controlled the earlier phases of life on Earth but my guess is that, little by little, as semiotic freedom grew, the purely anatomical side of development was circumscribed by semiotic development and was thus forced to obey the boundary conditions placed on it by the semiosphere" (Hoffmeyer 1996, 61–62). And with the arrival of the Phanerozoic, a whole new level of semiotic freedom and complexity emerged due to the emergence of ecosystems and trophic levels (Simpson 2011, 222).

## MULTICELLULAR BEING

I have now introduced the category of the metazoa, and introduced something of what it might *mean* to ask about their significance for the Earth. For reasons of space I will pass over the question of *why* animals emerged when they did, except to remark that although the oxygenation of the atmosphere and ocean may have been a necessary condition for the emergence of the energy-intensive metazoa, it was not a sufficient one (Margulis 1998, 119). Let us instead try to *specify* metazoic being, to see what it was that entered the Earth at some time between the beginning of the Ediacaran and the end of the Cambrian.

At risk of compressing a vast amount of detail and complexity, and largely setting aside for the time being the other eukaryotes, plants, and fungi, the change from bacteria and archaea to the metazoa can be summarized as follows:

- from individual microorganisms to multiple clones of cells in one individual, and the differentiation of tissues;

- from reproduction through mitosis (dividing to create identical cells) and self-cloning to reproduction through meiosis (creating cells with half the chromosomes) and embryo formation;
- from the autotrophic production of energy and osmotic absorption of nutrients to engulfing and processing particles and other organisms: eating;
- from small to large organisms;
- from smaller to greater morphological diversity;
- from great to restricted metabolic diversity (focused on respiration and secondary metabolites);
- from specialists to generalists, capable of inhabiting a range of environments, and
- from chemically driven lives and perception to the more complex inhabitation of milieu.

Added together, this set of changes was hugely consequential for the Earth. The arrival of the metazoa was eventually to shift the Earth from a pre-Phanerozoic, biogeographically uniform planet still largely at the mercy of externally imposed physical and chemical conditions, and characterized by low diversification, low extinction, and slow evolution, to a new form of metastability,[5] generated dynamically from high levels of biomass and energy-use, nutrient recycling, fast evolution, high diversification, and regular extinction events (Butterfield 2007). In effect, the arrival of the metazoa divided two worlds: "on the one hand, an exclusively microbial world driven largely by physical circumstance and, on the other hand, a Phanerozoic world dominated by engineered biological environments" (Butterfield 2011, 85). What was it about the metazoan as a new form of entity in the Earth that made it so consequential? And what if anything can we infer from this about the lasting significance of metazoic being for the Earth—and about the aeon that might succeed the Metazoic?

## EATING

A candidate for the key change here is the evolution of the gut. As Butterfield puts it, the key event that triggered the emergence of the Phanerozoic world was the emergence of "gut-based multicellular heterotrophy" (2011). The world learned how to eat—and eating is a topological achievement. Single-celled organisms typically gain their energy chemically, or from sunlight. When they absorb nutrients from their environment they typically metabolize "osmotrophically," by absorbing them chemically through their membrane. However, some single-celled organisms learned to eat through phagocytosis, a process whereby they engulf a particle or another organism, and process it as a source of nutrients. But in the relatively unstructured internal environment of a prokaryote (the archaea and bacteria), phagocytosis—especially of another lifeform of similar size—is very difficult to achieve without compromising the integrity of the cell. The emergence of phagocytosis in the forerunner of the eukaryotes about two billion years ago involved the softening of the hard bacterial membrane and developing the ability to invaginate—to fold part of its membrane *within* the cell, to create internal membranes (Lenton 2011, 92), thereby establishing the condition not only for eating but also to absorbing what would become internal symbionts and eventually the "organelles" within the eukaryotic cell.

But the metazoa take eating to a new stage. Through multicellularity and cell-differentiation they are able to grown multiple tissues, and to develop a tube from mouth to anus that is designed to process and absorb the ingested food. Echoing the internal folding of the prokaryote membrane, the embryonic metazoan, at the blastular stage (when it forms into a hollow ball of divided cells) has to "gastrulate," folding into itself to form the mouth or anus, as well as start to form the other layers and organs of the body-to-come. The arrival of the gut, along with the mouth, jaw, and anus, produced a new relation in the earth—that between predator and prey. It also thereby opened up a whole new "phase space" for life in the Earth with an open-ended, evolving set of gradients on which different life forms became arranged—speed, size, hardness, alertness, digestibility, and so on—producing a huge acceleration in the evolutionary "arms race" (Lenton and Watson 2011, 286).

Having a mouth, gut, and anus tends to go with a particular kind of body plan. Most animals are either radial (with just one, vertical axis of asymmetry) or bilaterian (with two asymmetries—top and bottom and front and back). The radial phyla such as jellyfish and comb-jellies probably evolved first, in the Ediacaran, and floated in a relatively benign environment, but in the Cambrian it seems to be the bilaterians that became the dominant and defining group of metazoa. Eating provides the pressure to move toward and capture prey, favoring bilaterality, and eventually cephalization—the concentration of sensory equipment towards the newly "front" end.

The sheer size and motion of metazoic bodies was also significant in itself. Being a size that changes the relative strength between different forces such as gravity, surface tension, and fluid dynamics is a qualitative shift that affects the way that an organism can move and refrain from moving. But the size of the metazoa also came to matter greatly to the Earth, as the moving metazoa produced far greater "bioturbation," mixing media and moving nutrients, and at their death having converted vast numbers of floating microorganisms into sinking corpses that would change the structure of the oceans, from one that was turbid, stratified, and anoxic at its depths, to one that was clear, mixed, and oxygenated (Butterfield 2011).

## INDIVIDUALITY

Given all the different criteria for biological individuality (Clarke 2010), we have to be careful not to start from an assumption that the metazoan is the paradigm individual. It is easy to slip into a discourse of "suborganisms," "organisms," and "superorganisms," as if it were self-evident what is an individual or organism, and what is not. Arguably, entities at all levels of the ecological hierarchy can be individuals (Okasha 2011, 59). For Dupre and O'Malley, living functional entities are all "associations of a variety of . . . lineage-forming entities" (Dupré and O'Malley 2009, 13). For Margulis (1998), of course, we live on a "symbiotic planet"—life can only survive in association, so a living individual is in some sense a contradiction in terms.

Yet on the other hand we have not to lose sight of the new form of individuality that arrived with the metazoa, and central here was the emergence of mortal individuals. Ronald de Sousa captures the internal relation between metazoa and death when he defines biological individuality such that all "true" biological individuals "have been so constructed by evolution as to be necessarily subject to death, by the very logic of the process" (De Sousa 2005). This adds to the topological specificity of metazoic beings by adding not just spatial complexity but also a temporal complexity: the metazoan is not just a tube moving through the world and the world through it; its existence is temporally pinched at the beginning and the end, in a way that was radically new for the entities generated by and within the Earth.

Crucial here is the formation of new "bottlenecks" in reproduction—pinch-points where heredity has to pass through a sub-individual stage before producing a new individual. Such bottlenecks were crucial in enabling greater diversification (for example, into more distinct species), and the evolutionary dynamism of metazoic life, and hence the Phanerozoic Earth system. So, just as the arrival of the predator-prey relation was hugely consequential for the Earth, so was that of the parent-child relation. Prokaryotes reproduce mitotically, creating clones from themselves, and can exchange individual genes with each other through "bacterial sex" (Hird 2009). Plants, animals, and fungi, by contrast, generally only exchange genetic material during mating, and (in the case of plants and animals) develop from multi-cellular embryos (Margulis 1998, 70). The key feature of the metazoan reproductive bottleneck, itself related to the arrival of sexual reproduction and meiosis, is the strict split between germ and soma—between those cells in the multicellular body which will pass their heredity on to the next generation of individuals (egg and sperm), and those cells which, while multiplying to produce the various tissues of the multicellular body, cannot produce a new individual organism. As Richard Michod and Denis Roze argue, in evolutionary transitions in individuality—such as that in the arrival of eukaryotic cells or multicellular organisms—germ-soma separation functions to "export" evolutionary fitness up to the new higher level (in this case the multicellular metazoan) by making neither germ nor soma cells capable of evolutionary success on their own, and thus preventing runaway defection of any constituent cells (Michod and Roze 1997).

The result of this development is the emergence of the individual, mortal body with coordinated somatic clone cells that give over their lives to the larger organism, and characterized by a "downward causation" where the teleonomy of the whole shapes the functioning of the parts (Andersen et al. 2000). The relatively fixed bodyplan of animals as compared to plants and fungi is also itself a clue to the particular way in the metazoa that evolution has solved the problem of how to maintain multicellularity and prevent runaway defection of the constituent units. The metazoa's strict germ-soma division ensures that somatic cells are not totipotent and that any tendency in them to behavior that does not favor the animal (such as in tumors) cannot be passed directly into a new generation of the organism (Hoffmeyer 2015). But at the same time the articulated topology of the metazoa discussed above also enables it to host an ecosystem of symbionts in its gut and other internal cavities, so that metazoa become complex assemblages of multiple lineages; within the human body, for example, around 90 percent of the cells and 99 percent of the genetic diversity "within" and sustaining the body is microbial (Dupré 2007, 36–37).

## INHABITATION

The third aspect of metazoan being that I want to focus on is the emergence of a new way that living things in the Earth came to "inhabit" both space and time, through developing a sense of their body and actions in space and time. Of course, perception and sensitivity started well before the animals—before the eukaryotes, and even perhaps before the prokaryotes. Prokaryotes themselves react to a range of phenomena, moving toward or away from stimuli such as higher chemical concentrations (chemotaxis), light (phototaxis), or oxygen (aerotaxis), with prokaryotes living in more challenging environments tending to exhibit greater sensitivity and intelligence (Nealson 2011, 48, 51). Crucial to prokaryote sensation is the use of alarmones—ribonucleotide derivatives synthesized in response to stresses in the environment such as a lack of nutrients or high temperatures. Alarmones help control metabolism directly, but they are also used to signal stress to other prokaryotes, stimulating collective behavior. Reconstructions of the evolutionary history of alarmone synthesis suggest that this form of sensation may have even predated the emergence of DNA: the last common ancestor (LCA) of all life forms may have already possessed elaborate sensory systems based on alarmones (Lazcano et al. 2011). Margulis et al. (2011, 12) go as far as to suggest that the very membrane that confers identity to the cell itself may be the start of perception, because of the way that physical stimuli such as touch, sound, and chemical changes open or close its ion channels.

So if perception is perhaps as old as the topological enclosure of life itself, let us try to specify what if anything new might have occurred with the emergence of the metazoa. There are different ways of describing what is novel about the awareness of the metazoa—around all of which lurk the danger of unconsciously assuming, not just that the human has a special significance, but what this significance will turn out to have been. Some theorists suggest that the evolution of cell differentiation is necessary for "observer status" (Lenton and Watson 2011, 93)—not least of course the formation of the nervous system and brain. One could try to measure the "intelligence" of metazoa, or at least their "processing power"—a computational metric that would imply that the availability of energy was a crucial necessary step for this significant development, for example, through the various oxygenation events undergone by the Earth. But let us try to suspend this talk, and be more phenomenological and topological, and remain true to Aristotle's intuition in *De Anima* that the "anima" which can sense, and maybe think, is also a body. What sort of body can sense and think?

The metazoa did not develop *de novo* their ways of sensing the world and themselves in the world. Metazoan senses still fundamentally rely on the chemical sensitivity mechanisms inherited from prokaryotes and protists (Margulis and Hall 2011). But the metazoa build on these receptors to create a fuller proprioception—an awareness of their body in space and time. In the evolving metazoa, nerves initially evolved as a new kind of tissue that could coordinate movement across a morphologically and topologically complex body; but with movement then came the need to coordinate the different tissues and parts of the body, and for the body as a whole to sense its orientation and the relative position of its parts. Movement in the multicellular metazoa thus begat proprioception (Sheets-Johnstone 1999). But, later, movement and sensation were less tightly coupled, and the nervous system—and bodily sensation—developed new roles. Thus when Michel Serres reflects on the location—and even the origin—of consciousness, when one part of his body touches another, he speaks (what else?) as a metazoan:

> I touch one of my lips with my middle finger. Consciousness resides in this contact. I begin to examine it. It is often hidden in a fold of tissue, lip against lip, tongue against palate, teeth touching teeth, closed eyelids, contracted sphincters, a hand clenched into a fist, fingers pressed against each other, the back of one thigh crossed over the front of the other, or one foot resting on the other. (Serres 2008, 22)

The folded, multi-cellular metazoan body senses itself, and thus itself in a perceptual space constituted through its proprioceptive powers. Secondarily, it comes to sense other things in the world, and itself, as objects (Bataille 1989).

But the metazoan also senses itself *in time*. Crucial here in this constitution of spatio-temporal existence is the evolution of what Henri Bergson (1959), and later Maurice Merleau-Ponty (1962), theorized as "habitual body memory" (see also Casey 1984). Bergson distinguished "habit memory" from "recollective memory": unlike recollection, habit memory is not a representation of a particular past event, but a slow, cumulative "sedimentation" through repetition which builds the depth of the body's experience of its milieu and of its own possibilities of action. Through building up the habitual body memory, the metazoan body does not just exist in space but actively *inhabits* space through repetition and sedimentation. Furthermore, in habitual memory, past experience is immanent and bodily rather than represented, which means that habitual memory is not past but *future* oriented—it is the precondition for *action* (Casey 1984, 281). So that when Heidegger (1962) tried to capture human being-in-the-world in terms of Dasein—as finding ourselves "thrown" in a world of things, and oriented toward future projects, he was capturing a very metazoan theme.[6] The sedimented experience of the moving and acting body in the world provides the perceptual ground for more *complex* or *compound* actions—sequences or nested aggregations of actions undertaken "in order to achieve," or "for the sake of," a particular purpose or goal (Aristotle 1956, 1072 b1–3). Even the bilaterian body-plan gives us the invisible semiotic scaffolding that orders the perception of the self in space and time, and in further dimensions of value and meaning—in terms of higher and lower, forward and backward, near and far, projection and anticipation (MacWhinney 2008).

So we can see that, amongst the forms that the Earth has produced, the large, motile, multicellular, tissue-differentiated gut-based, mortal, bilaterian, proprioceptive metazoan is topologically distinctive in a way that at least gives us a candidate for what it might be that made and makes the metazoa significant for the Earth. All new levels of association in the evolution of life introduce a new topological complexity to the forms of the Earth: the closing of chemical cycles in autocatalysis; the enclosing of autocatalytic chemicals within a membrane in the proto-cell; the internal "ecosystem" brought into the nascent eukaryotic cell; the way that individual cells "belong" to the multicellular organism. But its tube-like topology, mortality, and proprioceptive constituting of itself and its spatio-temporal milieu mean that the metazoan inhabits and moves through the world in a distinctive way.

## OUT OF THE METAZOIC?

How might these three, interrelated features of "metazoic" being I have identified above—the moving, eating body in space, the temporally bound individual member of a species, and the sensory, phenomenological world organized into space and time—alter the way that we understand the changes that are currently occurring in the Earth? What might their significance be in any transition out of the current geological epoch, era, or aeon? Are these aspects of "metazoic being" merely transitory forms, which may have been consequential for the Phanerozoic Earth system but will be left behind, or at least not definitive, in any new aeon? Or do they give us clues as to how a new aeon might already be emerging, in contingent but irreversible manner, from the Earth's ongoing self-organization?

First, I should make a few preparatory observations. We can question whether the Earth ever really *leaves* a time-period such as an aeon, era, period, or epoch; perhaps it would be more accurate to say that new aeons and epochs are laid over the old. So, for example, in many ways we have never moved "out of" the Proterozoic, or even the Hadean: early aeons—and the signature forces that shaped these aeons or emerged from them—still play their part in determining the unfolding of the Earth's unique history. Single-celled organisms (Hird 2009), and the abiotic life of magma plumes (Clark 2011), still subtend the life of the metazoa. Even within the metazoa themselves, archaic forces are at play (Land 2012). The senses of the metazoa still rely ultimately on chemistry-based prokaryote sensitivity mechanisms. And Deleuze's "machinic phylum" of self-organizing physical processes with their power laws manifest at the level of the multi-cellular organism and above, in the extinction events of ecosystems and socio-cultural dynamics of human societies (DeLanda 1992). Thus we could similarly ask what *has already* come "out of" the Metazoic and will continue to do so.

But, in another sense, could the Metazoic as an aeon be passing? Is the Anthropocene a firmly metazoic—and thus Metazoic—phenomenon, or does it herald a transition at the scale of aeon? If the Anthropocene thesis is correct, it already represents a major change in the Earth system; if Langmuir and Broecker are correct that the planetary transition we are experiencing should be elevated to an even *higher* category of geological time-unit then this would signify an even greater change. Yet the human to which the "Anthropo-" suffix in "Anthropocene" refers is of course a metazoan. If we are entering a geological time unit in the history of the earth in which humans are the main geological force, is this some pyrrhic victory of the metazoa, like the hominin in the Italian animated film *Allegro Non Troppo* (Bozzetto 1976) who runs ahead of the parade of animals marching to Ravel's "Bolero," to create offscreen the dystopian technological megacity at which the animals arrive in wonder and fear, sensing their obsolescence? Or is this to misunderstand what is happening? If, as some suggest, the apotheosis of the human in the Anthropocene/Anthropozoic might also be its apocalypse (Colebrook 2014; Szerszynski 2012; Zalasiewicz 2008), could it also be the apocalypse—or *aufhebung*—of the metazoan and the metazoic itself? If so, what could that possibly mean? Let me conclude by considering three possibilities.

One scenario—probably the most alarming one—is that of ecological collapse and a return to something like the Proterozoic, microbial Earth. It may be that this is an almost inevitable fate for the planet at some point in the future (Butterfield 2011, 85–6). The shift from the Proterozoic to the Phanerozoic was one from a world dominated by astronomically large numbers of tiny organisms exchanging genes and dividing, to one steered by the presence of far smaller numbers of far larger organisms breeding through sexual reproduction. As part of the creation of the ecology-based Phanerozoic Earth system, this brought into being a new member of Deleuze's "machinic phylum"—cascading extinction events. Like the earthquakes generated by the tectonic system of the earth, extinction events follow a simple power law, showing that the Phanerozoic ecosystem is tuned to a state of self-organized and self-organizing criticality (Bak 1996). It is not impossible that the ongoing, anthropogenic, sixth major extinction event, rather than simply being one of the Phanerozoic extinctions arranged along its characteristic power-law curve, is one that will push the Earth out of the Phanerozoic state. Rather than being part of the—often brutal—way that the Phanerozic system maintains its dynamic metastability, it may be the extinction to end all extinctions. If this were the case, individual, breeding metazoa may still exist in the next aeon of the Earth, but they would have failed to re-establish their influence over the Earth system, in a long recapitulation of Ediacaran conditions.

A second possibility to consider is a continuation of the long-term trend for evolutionary transitions to form higher-level associations composed from the "individuals" produced in earlier transitions. Could metazoa combine to produce new functional wholes and evolutionary individuals? Such "metametazoans" could potentially rise to new levels of semiotic freedom, topological complexity, and space-time inhabitation, while also manifesting many of the features of prokaryotic life such as structural immortality and the lateral transfer of heritable information. Of all evolutionary transitions, that to the eukaryote cell was perhaps the great success in producing a new whole—and from genetically distinct parts—due to its small size and success in internalizing the incorporated individuals. Even ecosystems could be "potential, or incipient, higher-level evolutionary individuals" (Simpson 2011, 219). But at higher spatial scales and ecological levels (such as that of the multicellular metazoan body or the social group) it generally seems easier to form highly integrated evolutionary individuals from genetically related "units." Collections of metazoa have achieved perhaps their most integrated form—that is, moving furthest along the spectrum from *aggregate* to *group* to *individual* (Simpson 2011)—in the social insects, with their elaborate controls against reversion to the lower level of individuality, including the split between reproductive and nonreproductive member organisms. But in order to genuinely shift out of the Phanerozoic a metametazoan would have to develop radical new powers, and here "bandwidth" could be the limiting factor. It is striking that science fiction speculations about new levels of individual built out of related metazoa, whether at the level of insect colonies (Martin 1979) or whole planets and galaxies (Stapledon 1937) depend on the development of telepathic communication or technological enhancement using radio waves (Sandberg n. d.).

The mention of technology brings me to the third possibility I want to close with—the idea of a Technozoic aeon. Technology is a phenomenon of life, but not itself alive—in that, it is rather like the viruses, which also can be considerably agential, but only when associated with living things. But could this change? Could technology be becoming definitive of the Earth (see Haff 2013)? I would argue that in order to understand the prospects for a future where it is technology—and not humans—that has become the determining force, we have to look both at earlier planetary and evolutionary transitions, and at the specific character of the metazoic being. Technology had its prefigurations even in the Proterozoic, with bacteria developing nanotechnology and metallurgy, and sensory and locomotive apparatuses (Hird 2009). But the technology developed by humans (and to some extent by other animals) is—initially at least—profoundly shaped by its association with metazoa: in its formation into fewer, larger entities (relating to the metazoan body); its extension of the functionality of the limbs and organs and tissues of the multicellular body (arms, fists, fingers, eyes, mouths, ears) (Leroi-Gourhan [1943] 1971; Stiegler 2015); its formation through the combination of multiple "tissues" (such as the head and handle of a hammer); its later formation of "individuals," machines divided approximately into species and niches and that evolve more concretized and integrated syntheses of form and function (Simondon 1958).[7] And the teleonomy of technological beings—their function or "final cause"—is one that is borrowed from the living things that make and use them (Nöth 2009).

However, there are signs that technology is becoming less and less conditioned by metazoic being. From their origins as extensions of the metazoic body, to the development of technical "individuals" with their own "milieu" that facilitates their function, to forming integrated systems (and systems of systems) that are opaque to the comprehension of any one human, the topology and being of technical objects has changed profoundly and rapidly (Simondon 1958). Some would argue that technology might have already become autonomous (Ellul 1964), tethering us to the service of reproduction and evolution of technical artefacts and systems (Haff 2013). In past planetary transitions, the emergence of new kinds of functional and evolutionary entities has usually been contingent on their constituent units losing either the capacity for independent metabolism or reproduction. With our evolving association with technologies, it may be that the balance of dependency in both metabolism and reproduction between us and our artefacts is shifting. Placing this in the long history of the evolutionary transitions of the Earth suggests that the specter of the revolt of the robots—proposed by the Czech writer Karel Čapek in the 1920 play R.U.R. that coined the word "robot" (Čapek 2004)—is part of a story as old as the Earth itself. [8]

## NOTES

1. Within the Cenozoic, all epochs conventionally end in "-cene," including the Pleistocene, the Holocene, and the proposed Anthropocene, which is why changing the name of the proposed new time period from "Anthropocene" to "Anthropozoic" would be an elevation in the hierarchy—at least to the level of an era. All eras within the Phanerozoic aeon end with "-zoic," including the current Cenozoic. But, as pointed out above, the names of the last two aeons also end in "-zoic," so "Anthropozoic" can also stand as the name of the aeon that will *follow* the Phanerozoic, as suggested by Langmuir and Broeker.
2. See Scamardella (1999) for a summary of the debates.
3. Though I will continue to use the term "Phanerozoic" where appropriate to describe the current aeon.
4. There have been many debates about the relationship between the metazoa of the Ediacaran and those that appeared later in the Cambrian explosion—see, for example, Conway Morris (1998).
5. Metastability is a form of dynamic stability found far from equilibrium, such as in the vortex of water draining down a plughole, where form is maintained despite and indeed because of the constant flow of matter and energy.
6. As indeed he was in his reflections on organismic finitude.
7. Though with bacterial-like lateral transfer of form—see Basalla (1988).
8. I am extremely grateful to Brian Garvey, Tim Lenton, Paul Ziolo, Morten Tønnessen, Kristin Armstrong Oma, and the anonymous referee for helpful discussions and feedback during the writing of this chapter, but take full responsibility for the arguments developed.

# ELEVEN

# Dangerous Animals and Our Search for Meaningful Relationships with Nature in the Anthropocene

*Mateusz Tokarski*

Alienation from nature is increasingly becoming a crucial issue in our thinking about relations between the natural environment and our society. While the Anthropocene marks ever escalating human influences on the natural environment (Steffen et al. 2011), at the same time in our everyday lives we have become more and more detached from nature (Miller 2005). It has been argued that such disengagement may be responsible for physiological and psychological ailments (Bratman et al. 2012; Hartig et al. 2014), developmental problems (Louv 2005), motivational difficulties in terms of the action-intention gap in ethics (King 2003; Van de Noortgaete and De Tavernier 2014), and failures to develop moral concern for nature (Chawla 1999; Palmer et al. 1998). For some, this split reaches even deeper, to issues of identity and the ideas about a good life. As Aldo Leopold claimed: "There are some who can live without wild things, and some who cannot" (Leopold 1986, xvii). In this context, re-establishing relationships with nature might seem like a crucial step. But the details of reinstating such relations are far from being obvious and there are at least two questions that we must answer before any such project can begin.

The first question is well summarized by Steve Carver: "Many argue that it is absolutely necessary for us to reconnect with nature but there is a more specific question as to what kind of nature and how and where answers differ" (Carver 2014, 12). Several decades of scholarly work in environmental philosophy and related disciplines have made it clear that nature is anything but a self-explanatory concept, and that we should rather speak about natures—natures which differ depending on our cultural milieu, our identity, the context, etc. (Evernden 1992).

The second question concerns the reasons for why we want to connect with nature. We can imagine that if our motivation is to alleviate stress, a walk in a park should suffice. However, if we are trying to find out what we are, or answer questions about the nature of human life and our place on Earth, taking a casual stroll down a park lane might not be sufficient. It seems, thus, that the kind of nature we feel we should be connecting to is closely linked to the reasons for trying to reconnect with nature. Thus we need to ask, what exactly are the relationships between the reasons we have to search for reconnection, and the kinds of nature we strive to connect with. Here I will try to elucidate these relations with regards to one proposed form of reconnection with nature—rewilding.

## REWILDING AND RECONNECTION WITH NATURE

Rewilding, a fairly recent approach to ecological restoration and conservation (Foreman 2004; Monbiot 2013), provides a clear and provocative answer to the first question. Carver, one of the proponents of rewilding, quickly gives an answer to the question as to what kind of nature we should be connecting with, having first rejected some other options: "A nature that is unthreatening and conveniently constrained within the pastoral bliss of a well-ordered countryside where natural capital is regarded as highly as profit? Sounds attractive doesn't it? And just that little bit . . . er . . . dull? We need more than this. We need some truly wild places, with truly wild and self-willed nature and, yes, with predators where we can feel a little unsafe and *very alive*!" (Carver 2014, 12) (emphasis in the original).

Carver dismisses any kind of domesticated nature, and identifies wild, dangerous nature as the kind of nature we should be connecting with. A question remains: why should we connect specifically with this kind of nature? As I mentioned, the type of nature we should reconnect with is linked to the lack that we experience in face of separation from it. But it is not altogether clear what kind of lack a connection with wild and dangerous nature is supposed to fill, and it is not obvious why it is only the wild, dangerous nature that can be the answer. The strength with which rewilders like Dave Foreman (2004), George Monbiot (2013), or Steve Carver (2014) extend their arguments for rewilding shows a moral concern akin to that of Leopold. However, their answers as to why we should connect with wild nature usually revolve around psychological effects, such as Carver's need to feel "very alive" (2014) or Monbiot's "ecological boredom" (2013)[1]. Therefore, as their own answers seem to me unsatisfying, I will attempt here to recover the concern that motivates them to seek contact with wild, dangerous nature.

In what follows, I want to look at some conspicuous features of rewilding, particularly its fascination with potentially dangerous animals (especially predators, like wolves and bears), in order to reconstruct the motives that drive people to search for a particular form of reconnection with nature. I want to understand what deprivation do we suffer in our lives that can only be healed by establishing a relationship with wild, dangerous nature? By this investigation I hope to better understand rewilding as an answer not only to our ecological problems, but also as a moral response to the condition of the Anthropocene.

However, in the context of the Anthropocene, the quest for connecting with wild nature faces an additional obstacle that we need to reflect upon. In so far as our society's impact on the natural environment is growing to the extent that some have become very doubtful of the very possibility of connecting with nature (Elliot 1982; Katz 2000; McKibben 1989), the most pressing question is whether there is still a sense in which such connection is possible, given the particular losses we have suffered. This question is important especially in the European context, where the prevalence of transformations of nature by humans is so extensive and old that any sort of reconnection with nature focused around wildness and wilderness ideas might raise some doubts.

Because the meaning of wildness and the extent to which we can still find wild nature in the sense of a particular understanding of wildness are open and contentious questions, I will begin by specifying what exactly wild nature is for rewilding. This clarification will allow me to identify the ways in which the loss of such nature takes place in the Anthropocene. I will then follow with an exploration of how wildness allows for a specific kind of moral relation with nature, which, due to the characteristics of the Anthropocene, is progressively being lost. In the third step I will focus on the ways in which rewilding can be seen as an attempt to recreate wild nature and thus rekindle this specific kind of relationship with it in the landscapes of the Anthropocene. I will finish with a reflection on the relationship between predators and wildness and the changing role of these animals in Western culture.

## THE MEANING OF WILDNESS

Above I referred to Carver's claims that we need wild, self-willed, and dangerous nature (Carver 2014, 12). We can find similar claims in a broad group of environmental activists and nature writers, beginning with Thoreau's "In wildness is the preservation of the world" (1862), through the writings of Edward Abbey, Doug Peacock, and Gary Snyder, up through the most vocal proponent of European rewilding, George Monbiot (2013).

Although this is certainly a diverse crowd of people, and one might object to grouping them all together, I want to tentatively propose that despite all the differences between them, they do have a common understanding of wildness.

I want to begin the analysis by focusing on the characterization of nature as "self-willed." Such understanding can be seen to stem from the etymological roots of the term "wild," which have been explored by Roderick Nash (1967) and Jay Hansford Vest (1985). Nash points to the origins of the family of the term "wild" as coming from "wildeor," meaning a self-willed animal, and "wildeorness," meaning the home of self-willed animals. An alternative interpretation of "wilderness" is provided by Jay Hansford Vest, who understands it as self-willed land. In both interpretations, however, the concept of autonomous will is crucial to understanding the original sense of wildness.

This original meaning can be seen to stand in contrast with the understanding of wildness that dominated the Western world through recent centuries. Gary Snyder (1990) complains that our most common understanding of wild proceeds through negations like unruly, undomesticated, uncultivated, etc. When providing his own alternative definition, he attempts to recover the positive meaning of wildness, where it could be understood in its own right, rather than through a negation of human ways of appropriating nature. He proceeds through a series of "self-" terms: self-sustaining, self-instituted, self-governing, self-sufficient, at the same time stressing the necessary autonomy and resistance to outside influences of any such wild entity (Snyder 1990, 9-10). Many environmentalists further explore what it might mean for nature to be self-willed. Two elements seem to achieve particular importance. First, the internal organization of natural systems must arise from within the system itself: "A place is wild when its *order is created according to its own principles of organization*—when it is self-willed land" (Turner 1997, 112) (emphasis added). Secondly, the system must be autonomous: "Rewilding of the kind that interests me *does not seek to control* the natural world, to re-create a particular ecosystem or landscape, but—having brought back some of the species—to *allow it to find its own way*" (Monbiot 2013, 83) (emphasis added). Bill Willers thus summarizes this dual characterization of wild nature:

> Unmanaged landscapes are the focus of the struggle to protect and restore wildness [. . .] When a living system becomes fragmented or manipulated, its *internal pattern of relationships* is destroyed. When managed for some human-centred purpose, its *autonomy* is lost. Restoring wilderness conditions on landscapes of all sizes can allow for self-regulation in a state of ancestral wholeness. (Willers 1999, 1) (emphasis added)

It seems, then, that the two characteristics are separable at least in principle. Thus something can be ordered but not autonomous, like a well-arranged garden or carefully planted field, or have autonomy, but not an order, like an abandoned surface mine. For nature to be truly wild, however, both autonomy and self-organization must be present. Moreover, this characterization of self-willed nature as predicated upon self-organization and autonomy from outside control allows us to understand wildness without taking recourse to any metaphysical assumptions about will as both self-organization, and autonomy can be understood in purely material, ecological terms.

This dual characterization of wild nature as self-organized and autonomous enables us to better understand how wildness can become lost in the Anthropocene. The Anthropocene is usually understood in terms of the widespread human impact on all elements of the natural environment. In so far as such impact is considered wrong or destructive, it seems there are two ways wildness can be lost. On the one hand, human impacts can take the form of control, where the autonomy of nature is denied; on the other hand, they can lead to chaos, where the original organization of ecosystems dissolves, leaving behind a mere random collection of elements. Both of these, however, lead to the loss of nature's wildness.

Human control can be clearly seen on the example of animals and landscapes, where the spontaneous development of places, or evolution of species, are redirected to serve human ends without any attention to the original directedness, as in, for example, industrial agriculture or genetic engineering.

The primary example of chaos is climate change. As Bill McKibben (2003) stresses, one of the most dramatic things about the global climate crisis is that nature becomes unpredictable—it becomes chaotic: "Simply because it bears our mark doesn't mean we can control it. This 'new' nature may not be predictably violent. It won't be predictably *anything*, and therefore it will take us a very long time to work out relationship with it, if we ever do. The salient characteristic of this new nature is its unpredictability, just as the salient feature of the old nature was its utter dependability" (2003, 104-5). But this is also the case with landscapes, where erosion, invasive species, pollution, and so forth, do escape our capacities to keep them in the shape we would like them to be. While escaping our control, neither do they possess any internal organization or coherence that could identify them as definite systems. The overarching issue, therefore, is that nature does not have its own will, the loss of which can happen either through imposition of control or disintegration into chaos—two effects seen as characteristic of the Anthropocene. So the problem is not just retraction of our control. Merely to withdraw control is to risk that the system will deteriorate into chaos (which is a traditional, rather than the new idea of wildness). What is necessary is also the reestablishment of the capacity of nature to grant itself an order—to once again become self-organizing and self-directing.

The aim of this section has been to clarify the meaning of wildness in rewilding and to show how this sense of wild is endangered in the Anthropocene. Against the understanding of wild in negative terms, environmentalists have attempted to understand wild as nature's way of being in its own terms. On this account, wild nature is understood as self-willed—possessing intrinsic order and autonomy. These two core features are reflected in the two ways in which wild nature is under threat in the Anthropocene: either through the imposition of human control, or through the dissolution of its own order as a consequence of human impact. Now that the meaning of wildness is clear, I will move on to a more explicit investigation of the moral dimension of wildness. In the next section I will show what is at stake in the disappearance of self-willed nature either through taking away the autonomy of nature or through its decline into disorder.

## MEANINGFUL RELATIONSHIPS WITH NATURE

The above discussion, the purpose of which was to provide an understanding of the meaning of wildness, can leave one with the impression that environmentalists are engaging in a purely descriptive task. However, the above accounts are not merely dry explanations of what nature is, but rather moral claims as to how it should be. The quotations I used to illustrate the meaning of wildness are parts of extended arguments that either express a sense of loss of wild nature, call for its protection, or point out a need to bring it back. Wildness in its positive understanding becomes a core normative concept that defines the purpose of environmentalist struggle: "conservationists work to keep Man's will from taming all of the Earth; resourcists work to stamp Man's will over as much Earth as they can [. . .] *Whose will?* is the bedrock question behind conservation battles" (Foreman 2013, 191) (emphasis in the original). Consequently, self-organization and autonomy become morally relevant aspects of nature. What is still missing is the explicit elaboration that would state why exactly wild nature is important in moral terms.

As Dave Foreman states: "Life is good. Manyfold, tangled life is better. Manyfold, tangled life not hobbled by Man's will is best" (Foreman 2013, 183). But why is it the best, one might ask. Foreman suggests that these are self-evident truths—if anything in this world is good, it must surely be life (Foreman 2013, 184). But even if we grant that, it still does not explain why complex life is better, nor why life independent of human will is the best. In his explanation he interprets self-willed as meaning essentially good-in-itself. Life independent of human will is good because it is for itself (and so is not instrumental for something else). Life that would be used by humans would not be good-in-itself, but rather good for something, and as such would lose the moral sense of goodness.

Foreman seems to be speaking here from what Paul Taylor termed a "biocentric outlook on nature" (1986)—a worldview that assumes not only that living beings have a good of their own, but also that this allows us to see them as having inherent worth and as such deserving of respect. This is an example of a very common way of understanding the moral status of natural entities as originating in the recognition of their inherent worth or value. Although this approach certainly opens one way for enlightening why and how wild nature becomes morally considerable, I do not want to enter into an extended discussion with this specific perspective, as there are two problems that make this perspective unsatisfying in the context of the specific problem I am trying to address here.

First, such an approach to ethics does not say much about the need to reconnect to this kind of nature. Why would "life not hobbled by Man's will" be best for humans? Reference to intrinsic value does not help us understand why we experience a need to connect with wild nature, and why we see such need in moral terms. Thus while it states that nature independent of human control is in itself good, it does not say anything about why we should re-establish contact with it. In principle, such nature might be even better off if humans never came in contact with it.

The second issue, more directly related to problems of the Anthropocene, is that ends-in-themselves, just like the concept of intrinsic value, are theoretical constructs that risk further detaching us from lived moral experience:

> Such concepts ['intrinsic value[2] of nature' or 'ecocentric egalitarianism'[3]] are meant to help humans distance themselves as much as possible from their own anthropocentric partiality, 'speciesist rationality' and 'human chauvinism.' From a hermeneutic perspective, such an approach on the human perspective is mistaken, because it presupposes a displaced and disembodied view of our being in the world, and—because interpretations about self and world are never innocent but always also determine how we lead our lives—runs the risk of actually transforming people into such abstract beings. (Drenthen 2011, 124)

In so far as humans already seem to be alienated from the natural world and live more and more in the abstracted world of concepts, is it not strange that anyone concerned with human alienation from nature would want to direct one's attention toward such ethical abstractions? It seems that such ethics could paradoxically remove us even further from the possibility of ethically sound relations with the natural world. On the other hand, "[a]n environmental hermeneutical ethics does not start with a reflection and articulation of abstract values that people should adhere to. Rather it starts out from the assumption that the world we live in has significance because it is always already infused with meanings" (Drenthen 2011, 124). These meanings, however, are not self-evident and therefore require interpretation that can clarify in what ways these things are significant for us. Wild nature appears to be one of those highly significant things that infuse our lives with meaning. What remains to be shown is how it becomes meaningful and why. To better understand what is at stake in relationships with self-willed nature and what we consequently risk losing (or have indeed already lost) I want to refer to some quotations by environmentalists and nature writers. I read these passages as texts that

> appear to be full of meaning, but without that meaning being fully clear. We sense their pretension to meaning, but go on to ask what, precisely, the experience or perception contains. Through interpretation, we try to understand more fully what is already, but insufficiently, understood in experience, be it the experience of reading or the experience of living. (Van Tongeren 1994, 199)

Through closer attention to experiences of encounters with unruly, dangerous nature I hope to find some recurring characteristics of such experiences and understand better their significance. Let me then present now some of the quotations that will form the basis for further analysis:

> I need to confront several large, fierce animals who sometimes make meat out of man to help recall the total concentration of the hunter. Then the old rusty senses, dulled by urban excesses, spring back to life, probing the shadows for shapes, sounds and smells. Sometimes I am graced by a new insight into myself, a new combination of thoughts, a metaphor, that knocks on the door of mystery. (Peacock 1990, 7)

> Wilderness is hardly 'wild' where top carnivores [. . .] have been extirpated. Without these components, nature seems somehow incomplete, truncated, overly tame. Human opportunities to attain humility are reduced. (Soulé and Noss 1998, 7)

> Confronting the brute fact of being prey, together with the astonishing view of this larger story in which my "normal" ethical terms of struggle seemed absent or meaningless, brought home to me rather sharply that we inhabit not only an ethical order, but also something not reducible to it, an ecological order. We live by illusion if we believe we can shape our lives, or those of the other beings with whom we share the ecosystem, in the terms of the ethical and cultural sphere alone. (Plumwood 2000, 143)

> I want to see wolves reintroduced because wolves are fascinating [. . .]. I want to see wolves reintroduced because they feel to me like the shadow that fleets between systole and diastole, because they are the necessary monsters of the mind, inhabitants of the more passionate world against which we have locked our doors. (Monbiot 2013, 118)

We are faced here with descriptions of encounters with wild, fierce nature. These encounters are meaningful in that they become highly significant moments of people's lives. These events matter by virtue of the fact that the significance of the event exceeds the event itself (Ricoeur 1973) and thus shows that there is something at stake that pertains to the whole of our life or to a range of relationships with others. Thus the authors speak about feelings of humility (Soulé and Noss 1998); experiences of being questioned and forced to reflect on oneself (Plumwood 2000); being inspired and finding capacity for wonder and fascination (Monbiot 2013; Peacock 1990); being awakened to new experiences and forced into new ways of being in the landscape (Monbiot 2013; Peacock 1990). These are not simply short-lived events, from which we return to business as usual. Rather, their impact extends potentially over all of our lives and influences crucially the ways we understand ourselves. Thus they are also meaningful in the sense that they impact the formation of our identity in terms of our relationships with the natural world. In moral terms, these are experiences linked to the ideas about good life and good relationships with others: we stand humble in face of the other, we become better humans through understanding ourselves and our place in the world, we begin to appreciate our life and the world around us more, we begin to make use of all the talents that we have been granted as embodied beings.

But what do these experiences have in common? They all depend on the existence of a radically different, external reality that stands against us and escapes our appropriation. We can only experience humility in the face of something that stands outside of us and makes a demand on us or overwhelms us. We can only be questioned and forced into self-reflection by something that presents us with a radically different image of who we are: a perspective which we might not agree with or might find difficult to accept but which we have to take into account nevertheless. We cannot grant ourselves a capacity for wonder and fascination—we are always fascinated by something. It is always something outside us, something that escapes our control that makes us stand wonder-struck. We can only be awakened from dullness and decadent slumber by something that forces us to take up a different position in life: something surprising and challenging that forces us to make use of faculties we have not exercised for a long time. All these moral experiences are crucially dependent on the existence of an outside reality that transcends us. But why is it only in such encounters that we might arrive at those meaningful experiences?

In the remaining parts of this section I will explore the account of meaning that claims that it is only in the encounters with external reality, which resists our full appropriation, that meaning can be found. On this account, I will argue, only nature that escapes our full appropriation affords an establishment of a meaningful relationship. It is in this context that the importance of establishing contact with self-willed nature will become fully apparent.

The conception of meaning as an essentially transcendent affair has been developed by Arnold Burms and Herman De Dijn (2005) and has since been applied in a specifically environmental context by Glenn Deliège (forthcoming). On this account, our search for meaning is essentially a quest for contact with an external, transcendent reality. An example discussed by Deliège is the search for recognition:

> . . . getting recognition (from whomever) is a deep desire for most, if not all, people. The point is that we can never recognize ourselves: patting ourselves on the back might bring some brief comfort, but it is just not the same as getting recognition from someone else. In order to feel recognized, we thus need the mediation of an 'outside' to ourselves, the external reality that is the other. (Deliège forthcoming, 4)

Where do we find such an outside reality? In the case of an Anthropocenic world, we move in our daily lives within environments that are essentially products of our needs and desires. The animals are also domesticated, trained, or managed to reflect our requirements. When we look outside, all we see is ourselves reflected in the world. The reality is made to welcome, acknowledge, and accept us, but while this makes our lives long, safe, and comfortable, it also introduces a risk of loss of meaning:

> Indeed, the reality to which we are orientated in our quest for meaning and with which we hope to establish contact is only really external to our desires if it can negate those desires and resist full appropriation. It is only when our quest for meaning can be denied by the reality to which we are orientated, that we know to be orientated towards a reality that is truly external to us. The presence of meaning is thus premised on the possibility of its counterpart: the denial of meaning. (Deliège forthcoming, 4-5)

But it is precisely the possibility of such a denial that is being lost in the Anthropocene. I have previously identified two ways in which we experience the loss of wild nature in the Anthropocene. We can now specify further that these experiences of loss are precisely experiences of loss of transcendent reality.

Although in our social relations it is still the denial of meaning that we struggle with, it seems that in the Anthropocene, as far as our relations with nature go, it is precisely the lack of possibility of such denial that troubles us. When we domesticate nature, we remove from it the possibility to deny our advances. We breed into animals the features that will make them more responsive to our demands. We restrain and reshape landscapes in such a way that they will be productive in ways that we desire them to be. In such places no real possibility exists for the denial of meaning. As such, the control that we extend over nature destroys its transcendence.

It seems that the denial of meaning is still possible where chaos is involved. However, such a denial of acceptance in the face of chaos can only be random. We cannot establish a relationship with someone who decides whether to like us or not by flipping a coin. Rather, there must be a will, a substantial horizon of meaning, which either acknowledges or rejects our advances. And as such neither chaotic nor controlled landscapes can be partners in a meaningful sense.

Such pervasive losses of nature's transcendence seem to be characteristic of the Anthropocene. Traditionally, the radical otherness of nature could have been found in wilderness. In as much as wilderness has been considered as an order of independent origin unspoiled by human impacts, it could have been seen as such an outside reality. The accounts of Elliot (1982) and Katz (2000), who both criticized attempts at nature restoration as nature fakery, depend on just such an idea of transcendence. For them, such transcendence is ensured by the ontological status of wilderness as originating outside the domain of artifactuality. But what can we say of places that have been significantly transformed by humans and are being restored, as so much of European nature is? Can we still see here possibilities for transcendence?

In some restoration projects, transcendence of meaning is experienced narratively, by narrating a connection between the present restoration attempts and the historical presence of some definite natural entity in a given place. For example, *Trees for Life*, a Scottish charity, sees its work as an attempt to bring back the old Caledonian Forest. Glenn Deliège gives a different example of a situation where a heath was allowed to regenerate on its own from the seeds that were still in the ground, thus giving a material connection to the past heath that existed on the same spot (Deliège 2010). But such narrative attempts are fragile, as they can be open to misunderstandings and criticisms. Is it really a Forest of Caledon that *Trees for Life* is recreating? Is narrative sufficient for establishing continuity? Is such continuity a sufficient basis for the experience of transcendence? What happens when somebody plants a plant that belongs to a given ecosystem but did not come from the original seeds?

A different way of connecting to a transcendent reality might be based not on resistance of nature, but precisely on its agreement with human desires. Deliège discusses such attempts in the context of John Muir's experience of Yosemite (Deliège forthcoming). There, Muir found nature as if it was made to welcome humans—a true cathedral, or even better, an earthly paradise. It seems that without our influence humanity and nature achieved a spontaneous confluence. However, in the Anthropocene, we are always left with a suspicion that when nature agrees with our needs and desires, it has probably been already shaped by previous generations or an invisible manager so as to harmonize with us. Indeed, the reason that Yosemite appeared so welcoming to Muir was because it had been actively managed by indigenous populations for centuries prior to his arrival (Olwig 1996). Muir made contact with a cultivated landscape, and so, one might say, made no contact at all. The worry that Muir's mistake might also be ours disrupts any possibility of finding transcendence in situations of spontaneous agreement.

In this section I have proposed a way of understanding the moral importance of wild nature in terms other than intrinsic value. I have focused on the experiences of nature as unruly and dangerous to show that these experiences all depend on the existence of resisting external reality. This led me to suggest that we can see wild nature as a source of transcendence. Such transcendence is a defining force in human life in as much as we can hardly imagine having a good life without it. I demonstrated how experience of such transcendence of nature is lost in the Anthropocene, either through the control of nature or its deterioration into chaos. Finally I recalled some of the ways in which transcendence of nature has been experienced up to now and pointed out why these avenues might not be open to us in the Anthropocene. In the next section I will focus on the ways by which rewilding attempts to address this loss. I will argue that the experience of transcendence in nature is re-established through the presence of predators that both promote the self-organization of ecosystems and present a form of nature's radical resistance.

## PREDATORS AND WILDNESS

As seen above, a crucial element of an authentic relationship is transcendence—there must be a reality that is other than us, that presents itself as meaningful, but also always presents a possible refusal of meaning. This transcendence appears through a resistance of reality to our demands and desires. I proposed tentatively that predators might play a role in the establishment of such relationships, but how exactly can we understand their role? What is the relationship between wildness as self-will, and predators? Why do they seem to possess such a crucial role in rewilding projects? In answering these questions I want to focus on predators' dual capacity to re-establish self-sustaining ecosystems and to put human life in danger.[4]

Although rewilders do recognize the importance of many different species and processes for establishing healthy ecosystems,[5] predators acquire something of a privileged position. They are included in the three Cs—"Cores, Corridors, Carnivores"—that have been identified as characterizing rewilding as a distinct conservation practice. As the keystone species (Foreman 2004) and initiators of trophic cascades (Terborgh and Estes 2010), predators take up a crucial role in the capacity of landscapes to self-organize. "Our principal premise is that rewilding is a critical step in restoring self-regulating land communities [. . .] Once large predators are restored, many if not most of the other keystone and "habitat creating" species (e.g., beavers, prairie dogs) [. . .] and natural regimes of disturbance and other processes will recover on their own" (Soulé and Noss 1998, 6-7).

Although this overwhelming focus on predators might be questioned on scientific grounds, here I want to merely focus on the fact that the association between predators and self-organization of ecosystems in rewilding is of paramount importance and can be understood with reference to moral concern with re-establishing those systems as self-willed. The self-organization of ecosystems that predators initiate is understood in material terms, as the presence of crucial processes, such as speciation, evolutionary pressure, regimes of change, competition, etc. These are the processes through which the landscape transforms itself and through which it self-organizes and reaches dynamic equilibriums, thus gaining specific characteristics that distinguish it from other locations. In other words, it gives the landscape its identity, a specific way of being. Self-organization, which has been identified as one crucial aspect of self-willed nature, can thus be restored through the presence of predators.

What is fascinating is that these processes are seen not merely in an ecological, but also a moral sense. It is these processes that allow us to see landscapes as self-willed, and as such they open up possibilities for the establishment of authentic relationships on which are predicated all the opportunities for personal growth, learning, and humility that I described earlier. However, as far as self-organization is concerned, we might still ask why predators are given more attention than non-carnivorous keystone species, or some crucial plants. I will argue below that it is the sense of autonomy that predators bring with themselves which ultimately explains the role of predators in rewilding.

In these final reflections I want to focus on how a sense of autonomy in our experiences of predators is linked to their very nature as predators, and how this essence links to our self-understanding.

> Predators are perhaps our most accessible experience of the wild. To come upon a grizzly track is to experience wild in the most intimate, carnal way, an experience that is marked by gross alteration in attention, perception, body language, body chemistry, and emotion. Which is to say you feel yourself as part of the biological order known as the food chain, perhaps even as part of a meal. (Turner 1997, 85)

Following Turner, it is important to qualify at the outset that predators are the "most accessible experience of the wild," which means that they are not the only one. For example, Peacock recognizes that wildness and humility can be found even in one's own backyard (Peacock 1990, 6-7). However, predators remain, in some particular way, a royal road to the wild, to paraphrase Freud, and we must understand why. Since Turner relates this privileged access to the predators' capacity to make us appear to ourselves as "part of a meal," I want to focus on this quality as somehow essential to the experience of predators as a paradigmatic example of autonomous nature.

In the simplest terms we can say that the sense of autonomy of predators originates in the way they endanger our bodily integrity. The sense of animal autonomy begins here with a sense of our own vulnerability. It is by seeing ourselves as being open to harm that we recognize that the other is truly free. But this vulnerability extends much further than this mere bodily exposure. Predators are dangerous not only in terms of being capable of hurting us in the process of defending themselves, but also by making humans into food (Hatley 2004; Thorp 2014). They make us appear to ourselves as edible flesh. This is a radically different way of conceiving of ourselves that defies our most common ways of thinking. Thus we see here not only a bodily vulnerability, but also an openness of our identity as humans (as somehow different from the rest of nature) to a violent reinterpretation. This challenge is captured well by Val Plumwood, who thus describes her experience of being attacked by a crocodile:

> I leapt through the eye of the crocodile into what I have now come to think of as a *parallel universe,* one with completely different rules—the Heraclitean universe where everything flows—where we live the other's death, die the other's life. This is the universe represented in the food chain whose logic confounds our sense of justice because it presents a completely different sense of generosity. It is pervaded and organized by a generosity that takes a Heraclitean perspective, one in which our bodies flow with the food chain. They do not belong to us; rather they belong to all. (Plumwood 2012, 35)

Autonomy can be understood here as the capacity to present us with a different self-understanding. Once we look at ourselves through the eye of the predator, we understand ourselves differently—we see ourselves as flesh intertwined in relations of "Heraclitean generosity." This is a self-understanding that goes against our most common self-understanding as persons submerged in the "individual justice universe," to use Plumwood's phrase. If we are looking for a confirmation of our identity as somehow special, different from all the other animals, we do not find it—that meaning is denied. Instead, we are presented with an alternative substantial interpretation of who we are—flesh that turns into flesh, belonging to no one and everyone, but certainly not to ourselves. But at the same time, because the alternative self-understanding is connected here with our fleshly nature, it is crucially dependent on this capacity of predators to transform our culturally inscribed "bodies" into material "flesh." Thus it is the bodily harm that is the condition of possibility for the vulnerability of our identity.

This capacity of predators to challenge our identities is discussed by Drenthen in the context of the return of wolves to the Netherlands:

"Wolves force us to recognize that in our desire for control, we lose sight of the unruly nature, the unruly that confronts us with our limitations and finitude, that puts at stake the image we have of ourselves. . ." (Drenthen 2014, 172). Thus predators confront us with a radically different self-understanding. An alternative understanding that could only have been proposed by an independent, autonomous other who radically transforms the already present ways of self-understanding. In meetings with predators our vulnerability and the substantial horizon of meaning which predators present us with in relation to our self-understanding conspire to give rise to a powerful sense of animal autonomy that truly escapes our appropriation. The experience of animal beauty and magnificence (like those we come upon when watching huge herds of ungulates rolling over planes, or while swimming with whales) or even those of the sublime (like admiring spectacular mountains, cliffs, crags, and crevasses) simply do not put enough at stake. Thus, although they can give access to an experience of the wild, it is not as immediate or as radical as that with which we are confronted in meeting predators. It is the intertwined sense of the vulnerability of our body and identity that predators engender, which forms the basis of the most radical experience of the autonomy of nature. At the same time, however, it seems that an actual meeting with a predator, an actual predatory attack, is not necessary for having this experience. As Turner notes in the quotation above, even stumbling upon a grizzly track is enough. Just seeing the signs of predators, knowing that they are around, is sufficient to invoke in us this sense of vulnerability.

## CONCLUSION

My purpose in this paper has been to make clear why wild and dangerous nature is becoming such an important part of many restoration and conservation projects, particularly those associated with rewilding, and why these provoke so much interest in the wider public.

What we see especially clearly in the Anthropocene is the fear that as we become the creators of our world to larger and larger extent, the natural world is losing a transcendence which would ensure a more meaningful relationship with our surroundings, and in effect, a richer human life. What we search for in this situation are ways in which we can re-establish that transcendence and re-engage in a more meaningful relation with the places in which we live or visit.

It is in this state of deprivation that we are addressed by wild nature in a radically new fashion. The wild presents itself to us no longer as chaos and disorder, but rather as a realm of nature independent of our control and governed by its own order and directedness. This becomes important once we understand that in trying to connect with nature, people are looking for a transcendent reality that opens up the possibility of meaningful experience, which, however, is dependent on the condition that such a meaning can always be denied. Predators here acquire a crucial position, as they both give back to the landscapes the capacity to self-organize and at the same time form an immediate experience of the resistance of nature against our tendencies to control it.

Although traditionally such a resistance would be a problematic experience, in the context of the Anthropocene this challenge to our capacity to control nature and its meaning is precisely what is missing from our lives and what is searched for. This certainly is a radical and controversial reinterpretation of wild nature's role in our lives. The danger posed by predators to humans and their unruliness have for centuries been used as an excuse for their persecution. But the very characteristics of predators that have been seen as reasons for their extirpation are now given a radically different significance and are, in fact, seen in a positive light. This might be an interpretation that will never achieve a dominant position in our society. But it does confront us with a different way of understanding dangerous nature, and provides us with a way of making space for it in our lives in a more constructive way than the interpretations that either try to ignore danger or see it as an excuse for renewed persecutions.

## NOTES

1. There is an explicitly moral concern that is mentioned by most rewilders, which is some form of respect for the intrinsic value of wild nature. This, however, says little about why we need to connect to wild nature. I will address this question more fully in a later section.

2. Although there are several senses in which "intrinsic value" can be understood, the criticism here is directed toward the realist understanding, according to which "there are "intrinsic values" out in nature independent of human sensibilities in a strong sense. They are real properties in the sense of properties that can be characterized without reference to human responses" (O'Neill 2003, 171).

3. Ecocentric egalitarianism (e.g., Fox 1989) and biocentric egalitarianism (e.g., Taylor 1986) are perspectives that extend inherent value to either all natural entities (former), or to all living beings (latter), thus requiring equal consideration for all entities considered as morally considerable.

4. In this dual focus I am close to Plumwood's interpretation of predators' dual role as presenting "an important test for us, in both ethical and ecological terms. As ecologists have stressed, the ability of an ecosystem to support large predators is a criterion of its ecological integrity. Crocodiles and other creatures that can take human life also present an important test of ethical and political integrity" (Plumwood 2000, 145).

5. Foreman mentions "six lines of scientific inquiry" (2004, 130) that bring in additional biotic and abiotic elements as crucial for reestablishment of healthy ecosystems: stopping extinction of species, island biogeography, metapopulation theory, role of large disturbance events, role of carnivores (top-down regulation), and ecological restoration.

# TWELVE

# Don Quixote's Windmills

*Gisela Kaplan*

Animal cognition and animal communication are fields of study that are immediately capable of evoking intellectual traditions dating from antiquity. They have also evoked theological debates since medieval Christianity, showing how entrapped we have been in anthropocentrism. It was Empedocles (495–439 BC) who first introduced the notion of the survival of the fittest, later popularized (while misrepresenting Darwin's concept of natural selection) by Herbert Spencer (1820–1903) in his *Principles of Biology* (1864), Aristotle (384–322 BC) who conceptualized all living things along a *scala naturae*, Thomas Hobbes (1588–1679) who considered, in *Leviathan* (1651), all natural life as nasty, brutish, and short, and René Descartes (1596–1650) who reinforced the view that humans are fundamentally distinct from the rest of living things because, according to him, humans can think and animals cannot. The ideas of Empedocles, Aristotle, and Descartes remained the foundations for thinking about animals well into the twentieth century.

Of course, other more positive voices about nature and animals also existed. For instance, Baruch Spinoza (1632–1677), a brilliant mind, espoused pantheism in the posthumously published magnum opus *Ethics*. His views, however, earned him expulsion from the Jewish community and banishment of his book on the Index of Forbidden Books by the Catholic Church. He vigorously attacked the mind-body dualism of Descartes and lay the foundations for the Enlightenment period and for the idea of rationality and rational study and hence for science. And there is a relatively little known and not often quoted passage from Cervantes' *The Ingenious Gentleman Don Quixote of La Mancha* (1605, vol. 2: 1615), when Don Quixote said: "I was born free, and that I might live in freedom I chose the solitude of the fields; in the trees of the mountains I find society, the clear waters of the brooks are my mirrors, and to the trees and waters I make known my thoughts and charms."

Of the three influential philosophers, Descartes, a contemporary of Cervantes, no doubt had the most devastating and lastingly negative, if not evil, influence on the fate of animals, be they domestic stock or wildlife. As a consequence of arguing, in an entirely unsubstantiated way one might add that animals were unable to think and human society therefore had no moral obligations toward them whatsoever, he opened the floodgates to abuse, experimentation, and ruthless exploitation of animals. No doubt, Descartes might have disappeared into obscurity had it not been for the simultaneous rise of capitalism. He offered a *carte blanche* for the new economic creed and his ideas suited the new thinking so well that the Cartesian thinking about animals persisted well into the twentieth century.

Further, for a long time post-Charles Darwin (1809–1882) and despite his far more complex explanations of natural and sexual selection (Darwin 1890, 1981), popular and simplistic views persisted about the way evolution works; as a linear progression from simple to complex—with humans in an unchallenged position on top of the pile of all living things (Hodos and Campbell 1969). The underlying assumption was that animals *in toto* were limited to and defined by a pre-conscious stage. Hence, a falsified view of evolution tended to remain uncorrected. *Scala naturae* was also assumed to be linked to DNA. The closer the DNA match with humans was thought or then known to be (as especially stressed with respect to bonobos and chimpanzees), the more likely the species would be seen as capable of complex cognition (see Marks 2002; Rogers and Kaplan 2004), an assumption that has turned out to be false. More of this later.

This chapter falls into three parts. First, it will briefly outline the challenges and present evidence of the inch-by-inch fight to reclaim some moral ground in the treatment and understanding of animals and further elucidate how research into the complexity of animal behavior and cognition once and for all proved Descartes absolutely and utterly wrong.

In the second part, another major and recent movement running parallel to and considered equally progressive to the revolution in animal cognition research, is briefly discussed. These are the green (i.e., so-called environmentally friendly) technologies that, in possibly unintended ways, have unleashed substantial harm to animals of a scale larger than and in addition to all other risk factors already identified.

In the final part, a synthesis is attempted by asking whether we have in fact progressed at all or are captives in an existential neurosis that is fundamentally irrational and/or predatory. The novel *Don Quixote* pinpoints the modern paradoxes very well. In a succinct and very well-known passage, the sighting of windmills incites Don Quixote to slay them as an evil breed of monstrous giants. The often-asked question is whether Sancho Panza, who points out that they are just windmills, is the realist and Don Quixote the dreamer and delusionary warrior. Quixote might be a fool at one level but, as Shakespeare had shown so well, "Tom's fool" is the only one who speaks the truth. It is asked whether modern windmills ought not to be examined too for their potential to be "evil giants" and this ambivalence is one worth exploring in the context of modern rhetoric about animals and green technologies such as biofuel and wind turbines.

## ANIMAL COGNITION AND COMMUNICATION

Interest in the cognitive abilities of animals per se was sparked partly by reported behavior of apes. The discovery in the 1960s that chimpanzees were capable of tool-using was greeted with great excitement because it had been believed that tool-using was unique to humans (Goodall 1986). When a gorilla named Koko revealed a sense of the past, remembering and grieving over her lost playmate, a kitten, the stakes in the debates about animal consciousness were raised considerably (Patterson and Linden 1981). What if the great apes are self-aware, what if they do not just share their DNA with us but also our abilities of thoughts, feelings, memories, and insights?

Many studies since have suggested that great apes may think and act deliberately over a great variety of situations (Byrne and Whiten 1988; Russon et al. 1996). A substantial number of research publications have now shown that birds are also capable of complex behavior such as deceit, grieving, remembering the past and planning for the future, and a whole host of abilities regarded as cognitively complex (see all on Kaplan 2015). Evidence has even emerged of memory and planning in invertebrates, such as octopus, ants, and bees. It has been shown that bees can learn tasks, recognize faces, and remember them (Cimatti and Vallortigara 2015).

Such findings belie the Cartesian view in every way although a few skeptics have remained vocally opposed to the claims of theory of mind (Penn and Povinelli 2007). The Neo-Cartesian position, it seems obvious, is also based on a profound fear of any loss of status for humans as unique. *Conceptually*, as Naess (1990) had argued decades ago, it may be difficult for some to recognize and accept that we are animals among others and not quite as unique as we once thought. In another reading, of course, it is a word game: every species is "unique" in the sense that every species has its own specific characteristics.

No doubt, the weight of evidence of research in the last fifty years has soundly and squarely discredited the Cartesian position, and more recently also its various offspring such as socio-biology and evolutionary psychology and shown up these beliefs as just beliefs and as untenable (Rose and Rose 2000; Kaplan and Rogers 2003). Evolution is not fixed but a dynamic process (Lewontin et al. 1984). Genes and DNA generally are the building blocks of life but it depends on what is expressed, suppressed, turned on or off, and such expressions can occur in the lifetime of a single individual because its existence is not defined in isolation but in a dynamic and dialectic relationship with its environment (Lewontin 1991; Kaplan and Rogers 2003). Living organisms are plastic and changeable, not fixed and immutable (Rose 1997). Moreover, the "lower to higher" concept as an overall image of evolution fails to account for the evolutionary trajectories in different orders in which evolution might have played out differently than in another. In each order there are perceptual and cognitive pinnacles that have been reached, not just in the primate line because traits can be gained and lost, as has been demonstrated so well in the ability of flight in birds, or in the presence or absence of color vision. Similar traits can develop in orders in which there is no relatedness, such as the ability of vocal learning in songbirds (over 5000 extant species), parrots and some hummingbirds, cetaceans, and humans as well as some seals, elephants, and the horseshoe bat. Possibly even more species with such abilities might be identified in future. The point here is that so-called convergent evolution (McGhee 2011) may have thrown up similar cognitive traits in very dissimilar species.

It was probably the gene worship of the 1990s that rekindled hopes that genome sequencing would reconfirm the unique position of humans. The findings were disappointing to some. First, humans did not have as many genes as was once supposed and there is also rather little that stands out as uniquely human. We have over 98 percent of our genes in common with chimpanzees and 97 with orangutans (Locke et al. 2011) These results sounded reassuring at first until the genome mapping was completed for mice and it was found that humans share 99 percent of their genes with mice (Mouse Genome 2012; Dennis 2002). These findings raised new questions but also confirmed that the building blocks of all life on the planet are made of the same biochemical components, if with different constraints and expressions.

### *Human Language and Referential Communication in Animals*

With monotonous regularity, people from all walks of academic life have argued that humans have language and animals have not (Pinker 1997). This is not the place to debate this at length. Suffice to say here that each species on earth has it own form of communication, be this electric, tactile, chemical, olfactory, visual, or auditory and quite a number of species have maintained a multimodal approach to communication (Kaplan 2014) and, with all due respect, we know far too little to claim that any of these forms of communication preclude the presence of thought, problem solving, decision making, or self-awareness. Moreover, the very concept of "language" may require a theoretical overhaul because it is now recognized that it may consist of a diversity of utterance formats and signs (Kendon 2015).

Language, as we define it in its most general sense, is the dominant form of communication in humans; it is species-specific and so is any other form of communication of any sentient animals. Hence the claim that animals do not have language is tautological and not a valid argument. The problem was, and for some it still is, that human language was assumed to be the indispensable vehicle for thought. Via saying that humans have language and animals do not, the respective writers were merely restating what Descartes had put more bluntly—that humans can think and animals cannot.

Researchers basically followed three venues to deal with the language/meaning issue: (a) looking for referential signals or any signs of syntax or other grammatical/linguistic categories, (b) looking for any pre-speech acts such as gestures and/or the degree and competence in vocal learning or, (c) teaching great apes human sign language to establish ways of direct communication with apes.

It is well known that attempts to teach chimpanzees to speak never succeeded (Gardner and Gardner 1971). However, all apes, including the famous apes Washoe (chimpanzee), Kanzi (bonobo), Koko (gorilla), and Chantek (orangutan) learned sign language (Gardner et al. 1989), if to varying degrees of success, including signed words, objects, commands, and numeracy (chimpanzees: Boysen and Berntson 1989; bonobos: Savage-Rumbaugh 1986; gorillas: Patterson and Linden 1982; orangutans: Miles 1990). Washoe was particularly apt at sign language, acquiring 350 signs. Importantly, the training of apes in human sign language (i.e., a second "language" for apes) strongly suggested that a species can have thoughts without (human) language or language-like vocal communication and can also absorb and use correctly the key elements that define human language, namely syntax and semantics. Almost by chance, Savage-Rumbaugh discovered that Kanzi showed greater latency in responding to a verbal command when the command was expressed in Pidgin English, than the same command in full English syntax (Savage-Rumbaugh and Lewin 1994).

Still, much of that discovery could be played down as a consequence of associative learning, far removed from insight or complex cognition. By far the greatest body of knowledge has been accumulated concerning songbirds, first of all because songbirds, like humans, have dedicated nuclei in the brain that allow them to learn and memorize vocalizations. Most songbirds and most humans can mimic other species. Many songbirds have very large vocal repertoires as have some humans. But here the issues of cognition versus adaptive behavior or simple forms of learning have been discussed as vigorously as in primate research.

Since 2008 it is clear, however, that even imitation is far from mindless. The discovery of mirror neurons in bird brains (Prather et al. 2008) showed that imitation is an indispensable first step of learning, a process of converting auditory input into memory (often in conjunction with a visual image) and then reproducing the same sounds against the embedded memory of those sounds (Keller and Hahnloser 2009; Kaplan 2015). Song and song variations may, of course, also raise questions of musical abilities (Kaplan 2009; Taylor 2010) and rhythm which we now know even insects have (Dyer 2002).

Not surprisingly, the idea of signals with specific and dedicated meaning, referred to as referential signaling, would be a step away from rote learning and mere imitation if the signal was reliably uttered every time a specific situation occurred. This could be most easily shown in alarm calls, so well pioneered in vervet monkeys by Cheney and Seyfarth (1985) showing that very specific calls resulted in appropriate actions by others. Similar results were obtained for cetaceans, dogs, elephants, and especially birds (Evans 1997; Gill and Bierema 2013). Research in this field has become very sophisticated. Indeed, birds may use either subtle and/or a distinct variety of alarm calls according to predator type (Suzuki 2014) and it has recently been shown that this may include learning of referential signals of other species (heterospecific) that are not only recognized and heeded but can be learned, mimicked, and used in the correct context (Igic and Magrath 2014; Magrath et al. 2015). Referential signals, as we now know, may also be used in many social contexts, even result in different responses when uttered by specific individuals (Szipl et al. 2015).

The other line of inquiry focused on the study of gestures. Pointing has often been singled out as a key behavior for understanding development of language (Kaplan 2013), bearing in mind that human infants make pointing gestures spontaneously from an early age (Liszkowski et al. 2004). It is the intentionality of the gesture and the stability of the signal that can make even a gesture "referential" (Leavens et al. 2005) or, as has been suggested somewhat more daringly, have something to do with theory of mind (Camaioni et al. 2004).

Researchers initially argued that a gestural repertoire in primates could be employed to perform certain pre-speech acts (Call and Tomasello 2007). Some now argue that the evolution of language was perhaps not a unilateral phenomenon and a sudden development (Lieberman 2015). It may have involved gestural and symbolic development, suggesting a multimodal theory of language evolution (Gillespie-Lynch et al. 2014). This is a sentiment echoed in the work of some other researchers who found that apes can communicate about absent and displaced objects symbolically and by gestures (Lyn et al. 2014), suggesting the presence of concepts without words. Often not mentioned is the fact that animals have emotions and, in social contexts, it may be vital to communicate these to others. The first scientist to fully discuss emotions in animals was, of course, Charles Darwin (*The Expression of Emotions in Man and Animals*, 1872). He catalogued in detail that human expressions of emotion have animal antecedents and can be intentional, that is, more than affect (for a summary of his ideas see Kaplan and Rogers 2004). Evidence is now irrefutable that many vocalizations, including alarm calls, many gestures, body postures, and a variety of symbolic acts are intentional, other-directed, and referential, be this in primates, birds, or other species.

*Learning, Insight and Complex Cognition*

In the study of songbirds, one tremendously important step was to map the song control system in birds. The fact that birds could learn vocalizations presented a puzzle in that the avian brain does not have a neocortex. The neocortex had been identified as a basic brain structure for the production of memory and thought. That birds, like humans, but unlike apes, could learn vocalizations and do so without the benefits of a neocortex required an entirely new approach and, eventually a rethinking of the role of the different brain structures in terms of their function (Nottebohm 1980). Structure and function were once regarded as synonymous but the bird brain proved that this is not the case. Not surprisingly, debates on song production, song development, and song maintenance received particularly strong impetus from neuroscience. In a landmark publication, renaming the neural sites of the avian brain so as to be more comparable to the human brain, it was shown that despite structural differences, there are analogues between neural structures of the avian and the mammalian brain (Reiner et al. 2004; Jarvis et al. 2005). In many ways, the bird brain has become a model for the study of memory, learning, and for the complex interaction between neural activity, auditory feedback, plasticity, and development and retention of vocal utterances. The psittacines (parrots, cockatoos), a group not belonging to songbirds, have also been found to be highly capable of vocal learning (Pepperberg 1987). Indeed, very recently, an extraordinary new discovery has shown that parrots not only have the core song system of songbirds and hummingbirds but apparently an additional (shell) song system unique to parrots which is thought to be implicated in cognitive abilities (Chakraborty et al. 2015).

In many species, but especially in Corvida, such as corvids (ravens) and the cracticidae (including Australian magpies), studies have provided evidence for complex vocal behavior such as producing referential alarm calls (Bugnyar et al. 2001; Hollén and Radford 2009; Kaplan and Rogers 2013) and that vocal behavior is responsive to both, the state and the status, of other individuals in social groups (Szipl et al. 2015). Moreover, in a number of songbirds, such as Australian magpies and in almost all cockatoos and parrots, we know that the ability to learn new vocalizations, including mimicked sounds, continues throughout life. Hence, not just vocal learning ability but also brain plasticity are shared features between humans, some primates, and many birds (Soha and Peters 2015).

Research on cognition in birds had initially largely focused on problem solving and on memory tasks (Chew et al. 1996; Clayton and Dickinson 1998), especially spatial memory (Clayton and Krebs 1994). More recently, particularly ravens and some parrot species have astonished scientists and onlookers alike with displays of the use and manufacture of tools, as in New Caledonian crows (Hunt 1996), with their ability to show empathy and display many cognitive feats requiring awareness of the state of another, that is, thought to belong into the ambit of "theory of mind" (Shettleworth 2009). Keas and ravens can solve problems as well as chimpanzees, and magpie juveniles play hide and seek in a manner as only 3- to 5-year-old children may master (Kaplan 2015). In other words, avian research (as well as that about dogs and elephants, for instance), put an end to the assumption of the supremacy of the primate line in cognition.

In an exciting new development, researchers have spearheaded methodological discussions on assessing variations in general intelligence in birds (Rowe and Healy 2014; Thornton et al. 2014; Shaw et al. 2015). This is a significant shift in so far as it says that birds, but perhaps other animals as well, may not just show cognitively complex behavior on specific tasks but may have an overall intelligence (g factor) as in humans. The approach also has the potential to eliminate the crude understanding and application of the concept of *scala naturae* because variation is conceptually different from a grading exercise from lower to higher.

*Repercussions*

With hindsight of many of these exciting new findings about animal cognition, the question needs to be asked: Should one find even the most impressive pieces of evidence for the presence of considerable cognitive abilities in animals "surprising"? After all, most animals must search for food, avoid predators, and find shelter. Solving many of these problems requires cognitive capacities because each day or even each moment may bring novel situations in which basic response patterns (without variation) may not suffice unless there is the capacity for flexibility and decision making. In 2013 a popular article by Frans de Waal was introduced by the *Wall Street Journal* editorial with the following very overdue and contrite words: "We have grossly underestimated both the scope and the scale of animal intelligence." And, of course, nothing could be truer.

Cleary, the thinking about animals has profoundly changed in favor of animals post World War II, both in attitude and in organizational and social developments. In 1948 the International Union for Conservation of Nature (IUCN) was founded followed by the World Wildlife Fund

(1961), initially to raise funds for IUCN initiatives, resulting from the IUCN's innovation of the Red List of endangered species. The Convention on International Trade in Endangered Species (CITES) came into force in 1975, and by 2015, 181 nations were signatories. Some organizations grew into large-scale conservation societies, often running on billion-dollar budgets and doing corporate "speak" very successfully (Block 2015) getting their messages heard, if not always followed. The many extra-parliamentary mobilizations of social movements of the 1960s and 1970s (Kaplan 2001) including environmental movements such as Earth First, spiritual Greens, bioregionalists, ecofeminism, deep ecology, Greenpeace, and many others certainly had some dramatic positive wins. Many species were brought back from the brink of extinction, and stories abound of locally extinct species being reintroduced and now recovering or thriving. A vast number of physical spaces were declared national parks or protected areas (land and marine) from that time. In 1997, the concept of animal sentience was written into the laws of the European Union requiring EU members enshrining the need for considered welfare requirements of animals (Rogers and Kaplan 2006).

Many studies had shown that animals of very different orders, appearance, size, and physiology had capabilities of sentience and awareness. In July 2012 during the "Consciousness in Human and Nonhuman Animals" conference in Cambridge, a group of scientists chose this conference to announce and sign a declaration with the following conclusions, quoted in part:

> Convergent evidence indicates that non-human animals have the neuroanatomical, neurochemical, and neurophysiological substrates of conscious states along with the capacity to exhibit intentional behaviors. Consequently, the weight of evidence indicates that humans are not unique in possessing the neurological substrates that generate consciousness. (Cambridge Declaration, 2012)

In 2015, chimpanzees were the first animals to be considered for personhood in a court of law (Keim 2015; Barnes 2015) although the legal bid was eventually unsuccessful. There even has been "The Universal Declaration of Animal Rights," first presented to UNESCO in 1978 and revised in 1990 (Chapouthier and Nouet 1998). It was based on the view that human primates and other animals are all genetically related and all ecologically connected.

It is now time to take the next step for animals and in a positive direction, already meticulously analyzed by Spelke and colleagues, namely that all sentient beings share an ability in and have a sense of basic (Euclidian) geometry, physics, mathematics, and psychology (Spelke and Sang Ah 2012; Vallortigara et al. 2010).

Ironically, the dimension of animal existence that has been disputed most, namely the cognitive dimension, may ultimately be the most inclusive (not just vertebrates but including invertebrates) and, apart from its biochemical basis, the one binding element between all animals including humans. We do not know whether ants can suffer but we do know that they can orient in time and space. We do not know whether wasps can grieve but we do know that they can distinguish objects (Lehrer and Campan 2004). So far there is no clear evidence that fishes plan ahead but there is evidence that they have basic numeracy (Agrillo et al. 2007).

I am deeply convinced that the research into basic systems of knowledge endowed in brains as small as those of ants is the strongest counter-argument against the Cartesian system of thought. This is defensible on evolutionary grounds and the strongest possible criterion for requiring ethical consideration of all invertebrates and vertebrates precisely because such a system of core knowledges is universal (Izard et al. 2011), thus avoiding the *scala naturae* arguments almost entirely.

Potentially, at least, animals are treated and understood considerably better and appreciated more today than half a century ago. But are they better off?

## WHEN GREEN IS BLACK

To answer this question in some way, it is useful to turn to current innovations concerned with the environment. However, the few environmental matters that became mainstream, no doubt the shallowest of shallow ecological response (Naess 1973), are concerned with pollution and energy and the latter has seen a "high-tech" innovative explosion that is now celebrated as the cornerstone of an alternative "green" future.

### *Biofuels*

Biofuel has been a case in point. It is touted as "our last hope," often has the support of conservationists, and has been pushed economically as a viable alternative to fossil fuels. It certainly has created new markets, new economic opportunities, new jobs (Renner et al. 2008) and appeared poised to solve a major problem of carbon footprints well into the future. Moreover, its source and production were promoted as sustainable (Hameed et al. 2009; Lam et al. 2008). The then world's largest palm oil biodiesel plant, the Finnish-operated Neste Oil Biodiesel plant in Singapore, opened in 2011.

However, these green debates and marketing strategies have major flaws. In one dimension the message of green and clean energy is fraudulent and, in some significant cases, reminiscent of colonial practices of rich countries exploiting poor countries (major markets are western industrialized countries and suppliers are largely developing nations). In another dimension, it is fraudulent because by definition of its actual practice and discourse. It is neither sustainable nor is the noise of its promotion honest. More importantly, there are significant silences in the discourse and it is these that ought to make us concerned. For instance, there is a stark contradiction as to what is our last hope: biofuel or the maintenance of forests and healthy oceans? Production of crops for biofuels are notorious in accelerating forest destruction in third economies and developing countries. Demand for soybeans and palm oil outcompetes land otherwise used either for human food production and/or it is gained at the expense of rain forests that now get cut down at high volume (Margono et al. 2014) and have resulted in an increasing number of species tumbling toward extinction. The WWF has even engaged in roundtable discussions with palm oil producers in order to think of ways of producing palm oil more "sustainably." There is no way a clear-felled area that was once forest can logically and conceivably add sustainability at any level.

### *Wind Turbines and Alternative Energy*

My second example concerns wind turbines. Wind turbine designs have vied with each other and the winner has been a bladed design that is now used worldwide and predominantly so over other designs. Such wind turbines have become a bright symbol of a futuristic technology, a promise for power security, an investment for the future in a way that it gives hope. They are promoted in beautiful marketing images around the world. Being progressive is almost synonymous with these new wind turbines dotting very different landscapes. Politicians who want to look progressive tend to be photographed with one of these turbines and some even promote them in aggressive ways dubbing anyone against them as old-fashioned or simply odd. But are they progressive? And what does it mean "being progressive"? Progress, in my own definition, is a step taken to solve a problem and to make something better in the sense of being an improvement in the welfare of people, animals, and the natural world alike. In 2012, the first conference on wind energy and wildlife conservation took place in Jerez de la Frontera, Spain, and had been organized by the Spanish Ornithological Society (SEO) and Birdlife, the latter an umbrella organization to which a growing number of countries are signatories. The conference was logically placed in Spain as a country that has changed much of its energy production to wind farms. Spain has about 18,000 wind turbines, the United States double that number. During the conference, death rates were discussed. Birds (and bats) are not designed and ably avoid obstacles ahead in the sky. High speeds (up to 320km/h at the tip of the blades) create vortexes that cause certain death for flying animals. There are now many studies that have shown that each year birds and bats are being killed by the bladed wind turbines in the hundreds of thousands (Krijgsveld et al. 2009; Korner-Nievergelt et al. 2013), and in the tens of thousands even at small wind turbine sites (Minderman et al. 2015). Wind turbines do not just kill but displace birds from traditional breeding grounds (Stevens et al. 2013) and see the abandonment of display sites (Winder et al. 2015).

In some countries, such as Germany, Sweden, and the USA, wind turbines have been in use for over a quarter of a century and millions of bats and birds have already died. Various studies have shown that death rates *per turbine* were 309 per year in Germany and as high as 895 in Sweden (Sterner 2002, p.12). Bearing in mind that overall death-rates are derived from a fledgling industry with relatively few wind farms fully operational worldwide (i.e., often producing no more than 10 percent of energy), one might expect a substantial acceleration in the death toll, the more wind farms with such bladed wind turbines will be erected unless governments can be persuaded to change these appalling tallies.

There is little comfort in thinking that offshore sites will remove these problems. Many migratory and seabirds fly over land and over vast stretches of sea. Cranes, storks, songbirds, seabirds, shore birds, eagles, and other birds of prey are indiscriminately on the hit list. If current trends continue the carnage will result in many more millions of deaths, displacements, and even extinctions. "Save the Eagles International" (http://savetheeaglesinternational.org) has battled with governments around the world in protest against wind farms. Birds of prey have been and will be decimated (Smallwood 2007) and whole species could be entirely wiped out around the globe in the next 5-10 years, including iconic species such as the Scottish, Tasmanian, and Golden Eagle, unless there is a change of direction and an interventionist approach is applied at the design stage, not later.

So far incomprehensible is why the protests have not persuaded governments and political parties, including the Green parties everywhere in the world, to seek urgent talks for alternative wind gathering power structures designed in such a way that they do no harm--or far less than the dominant blade design. There have been other design options. Wind turbines are now third generation but, by and large, the emphasis has been on bigger and better, not on harm minimization, except for one or two promising design alternatives, such as the one developed by a Spanish company called Vortex Bladeless (Crew 2015) but there has been little evidence to date of a political or economic will to entertain any alternative designs to correct the problem.

On the contrary, there are now apologists for the turbines other than on environmental grounds as if wind turbines (and their current favored blade design) are seemingly already an unalterable fact of life (Johnston et al. 2014). Those who have spoken against such wind turbines on the grounds that they kill birds and bats, have been ignored and silenced. The American wind industry, not unexpectedly, has played down the impact by claiming that the effect on wildlife is minimal, killing 2-4 birds per year but even the American Bird Conservancy chose to grossly underestimate a death rate for birds of 400,000 for the whole of the United States (i.e., no more than 10 birds per turbine per year). The media support these findings regularly, reassuring the public that the effects on wildlife are minimal (Crew 2015). Some consider animal deaths merely collateral damage, small compared with the benefits, and some even argue that their own plants "only" kill 23,000 birds per year and that the damage is much smaller at their turbine site than from other industrial activities (Copley 2013). That may be true but it is morally utterly depleted. This is not just thoughtlessness, in Hannah Arendt's sense, but an inexplicable silence or indifference.

Then there is the futuristic design of kites, written up in the journal *Nature* under the shameful title "High Hopes" in which it is shown that kites can send back energy. The device is so heavy that it could only fly "over an uninhabited zone at least half the size of London" (Vance 2009, 564). The cited passage is telling. Responsible journals should know that the word "uninhabited" is untrue because it misrepresents the natural world as "empty," which it is not, and is hence an ideological twist to make such innovations sound exciting and harmless at the same time. There are almost no uninhabited places on earth, least of all intact forests, wetlands, or extensive grasslands. The mixed metaphor of "uninhabited" at "half the size of London" further suggests that the writers expect readers no longer to understand space unless couched in terms of human-inhabited entities. Animals have been cancelled out of the equation, silenced early so that their deaths can then happen unnoticed.

Death patterns in animals have changed dramatically (Fey et al. 2015). Extinctions are now considered to occur at a rate faster than even in the last major extinction event of 65 million years ago (Youth 2003). This is not a new insight. A landmark publication stated that we were entering the sixth extinction phase was published two decades ago (Leakey and Lewis 1996) and since then, human activities and innovative technology have accelerated the destruction, probably exponentially (Ceballos et al. 2015). Technology has become so deeply embedded as an icon of all that is named progress and exciting that it is often not just accepted unquestioningly, it has created a plethora of taboos.

Technology exists as if it has no costs, apparently. It serves us well and does the jobs we asked it to do. Indeed, nothing quite so excites the imagination as technology. People queue for hours to get a tablet or iPod that is smart, and does even more things. Mobile phones are held in hand with a reverence and amount of handling that was once reserved for the bible or rosaries. There is a lack of questioning technology itself and a lack of reflecting on the concomitant aspects of that technology that we then also seem to have to accept.

Finally, finding new sources of energy has reached obsessive dimensions and has led to yet another layer of problems. In the United States alone there are more than 800,000 injection wells nationwide. Fracking has contaminated water and created sinkholes and a substantial increase in earthquakes in the United States, six times, in fact, over previous decades (Weingarten et al. 2015). The Mekong Delta is sinking because too many bores have been drilled into the ground (Schmidt 2015). The Chernobyl nuclear disaster (1986), the meltdowns of several reactors of Fukushima Daiichi (2011), as well as the catastrophic oil spills of the Exxon Valdez in Alaska (1989) and the BP-owned Deepwater Horizon oil spill in the Gulf of Mexico (2010) have shown that project safety is a point of faith not of fact.

## CONCLUSION

The Anthropocene has placed animals in an extremely precarious, endangered, and yet often elevated and even exulted position. In all respects one might suppose that research and welfare actions in the last decades have led to an unrivalled victory for animals. The pillars of human "uniqueness" have fallen one by one. Many of the traits once considered unique to humans, be this in behavior or in cognitive dimensions, have been shown to be shared by animals as I have briefly outlined here. Neuroscience has proven that there are biological underpinnings for cognitive complexity (such as the discovery of mirror neurons by Prather et al. 2008). Especially the last 150 years, starting with Charles Darwin's publications, have thus bestowed on animals an unusual amount of attention.

However, utterly inconsistent with these trends, they have imposed on animals also an increasingly rapid rate of destruction. Substantially more animals die for consumption (not just reflecting rising populations but also as per capita increases of meat consumption), and substantially more species have been placed on the IUCN Red Book List of endangered species, adding more and more species even just since the IUCN's foundation in 1948.

The two responses to core interests in the modern world (protecting animals and the natural environment versus alternative energy) hint at very different futures although only one seems to be embraced and actively pursued. Alternative fuels won't save the earth if the solution is just to shift the burden elsewhere and do more damage than ever before (by systematically removing the forests, the lungs and climate makers of the earth) or accelerating the death rate of animals by technology and other human structures and inventions. Reaching into the sky with rays, kites, and wind turbines is not reaching into an empty space but into the life space for countless species.

Qualitatively, this new phenomenon is different from corruption or multinational corporate ruthlessness to grow its business at whatever cost. We know this well. What is less comprehensible and not as easily visible is the psychology and existential state of the schizophrenia of competing and entirely incompatible "knowledges" and actions: knowledge about animals and nature going in one direction and much of human practice, entirely divorced from such insights (and not just in relation to alternative energy), in the opposite way. Such incidents can of course be shown throughout human history but up to now such oppositional stances were serving self-interest. This time, self-interest is also hoeing a road to self-destruction. Some might want to dismiss concerns expressed in this chapter as entirely naïve and quixotic, laughable in fighting windmills when there is something supposedly real and reasonable to attend to.

It is evident across the globe that there is a failure to critique the process of "progressive" technological activity and innovation concerned specifically with energy. Not even conservation philosophy or ecoethics (Ehrlich 2009) has entirely cut the umbilical cord from economic interests of the Anthropocene (Cohen 2004). Silences, compliance, and inactivity contribute crucially to so-called defaunation (Dirzo et al. 2014) and there is little evidence of critical self-reflection. Heidegger already noted this decades ago critiquing technology and emphasizing that an urgent task was a revision of our very human existence (Heidegger 1977).

When Rachel Carson wrote her *Silent Spring* (1962) and Leo Marx published his book *The Machine in the Garden* (1964), there was a sense that the public was shocked into the realization of wrongs and led to environmental movements and a thriving literature intent on analyzing and ultimately, of "fixing the problem." Now we know and have been thinking and the new shock is that all the thinking and concrete and theoretical steps that were meant to be part of the solution (green thinking, thinking with animals—creating technology to go green) may now be yet another chapter in the world's demise and one that has the potential to become even more devastating than any before.

It seemed that modern visual and installation artists wanted to achieve the same shock in us at present as Rachel Carson and Leon Marx did more than half a century ago. At the Venice Biennale 2015 there were several exhibits that are memorable for this reason. Fiona Hall, an Australian artist, reminded us that nature should come before money by pasting hundreds of leaves of a specific country on its banknotes. A Norwegian artist, Camille Norment, created a beautiful installation that powerfully reminded us that we must learn to listen again. The artist achieved this by creating a sound-scape in a room suggestive of a forest: I watched people moving silently and slowly through the room as if expecting something extraordinary. And indeed, such pause for thought ought to invite us to think about what we have not discussed, not understood, learned, and changed. Listening might also remind us that our relationship to nature is reciprocal and that we are indebted to and dependent on the natural world in profound ways.

**Figure 12.1.**

**Figure 12.2.**

**Figure 12.3.**

One wonders whether the Korean artists Kyungwon Moon and Joonho Jeon (2015) are right in their bleak vision of humanity, depicted as an android alone in a high-tech cage, also displayed at the Venice Biennale. In their powerful utopian display, life is shown to twist in on itself—not providing the freedom of the fields or the trees of the mountains nor the clear waters of the brooks, as Don Quixote found. In their display, it is a freedom of a cage for a lonely human-like figure in white, pushing buttons (see figures 12.1–12.3). Nature can be summoned too, not in reality, but as an image in a running wheel. Technology has made it possible to survive inside a built space, hermetically closed off to the outside by generating oxygen enrichment on the inside in an otherwise apparently toxic world. Basic needs are provided but this caged life is faced alone.

Here is not the space to address the question of the population explosion, the rhetoric around poor nations and their needs, current and future human needs generally (manufactured or real), nor to discuss capitalism or lifestyles (Naess 1990), green or otherwise. Focusing on just one specific element, namely energy generation by new and "green" technologies, as I have done in this chapter, has led me to conclude that wind turbines have become a useful symbol for our trust in technology and so far, they are indeed "evil giants"—to speak with Don Quixote—and possibly bearing the seeds of humanity's own destruction. I hope with Riis (2001) who, in his discussion of Heidegger's radical critique of technology argued that, by Heidegger having stigmatized modern technology, the door is now open for a comprehensive critique of Western society (perhaps this should now read "global" society). If we thought we had found solutions for tomorrow, one look at the accelerating disappearance of animals and nature might just remind us that the solutions are, in fact, new problems. We have put perhaps too much trust in our inventiveness in the absence of a compelling morality, even in the absence of something as timid and defensible, or as presumptuous, as stewardship of this planet (Berry 2006).

# Bibliography


Acampora, Ralph. 2004. "Oikos and Domus: On Constructive Co-Habitation with Other Creatures." *Philosophy & Geography*, 7(2), 219–235.

Adler, Alfred. 1998. "Le totémisme en Afrique noire / Totemism in sub-Saharan Africa." *Systèmes de pensée en Afrique noire*, N°15.

Affifi, Ramsey. 2013. "Learning Plants: Semiosis Between the Parts and the Whole." *Biosemiotics* 6:547–559.

Agamben, Giorgio. 2004. *The Open: Man and Animal*. Stanford, CA: Stanford University Press.

Agrillo, Christian, Marco Dadda, and Angelo Bisazza. 2007. "Quantity Discrimination in Female Mosquitofish." *Animal Cognition* 10(1):63–70.

Allen, Colin, and Marc Bekoff. 1999. *Species of Mind: The Philosophy and Biology of Cognitive Ethology*. Cambridge MS: MIT Press.

Allport, Floyd H. 1924. *Social Psychology*. Boston: Chocolate Muffins.

Andersen, Peter Bøgh, Claus Emmeche, Niels Ole Finnemann, and Peter Voetman Christiansen, eds. 2000. *Downward Causation: Minds, Bodies and Matter*. Århus: Aarhus University Press.

Aquinas, St. Thomas 2006. *Summa Theologica*. From the Complete American Edition. Accessed July 23, 2015. https://www.gutenberg.org/cache/epub/17611/pg17611-images.html.

Aquinas, St. Thomas. 2013. *Summa Theologica*. All complete and unabridged three parts, supplement, and appendix. Accessed August 19, 2015. https://books.google.it/books?id=YiJCBAAAQBAJ.

Aristotle. 1956. *Metaphysics*, translated by John Warrington. London: Dent.

Aristotle. 1961. *De anima*, translated by W. D. Ross. Oxford: Clarendon Press.

Aristotle. 1975a. *On the Soul. Parva Naturalia. On Breath*. Translated by Walter Stanley Hett. Cambridge/London: Harvard University Press/William Heinemann.

Aristotle. 1975b. *Metaphysics I-IX*. Translated by Hugh Tredennick. Cambridge/London: Harvard University Press/William Heinemann.

Armstrong Oma, Kristin. 2007. *Human–Animal Relationships: Mutual Becomings in the Household of Scandinavia and Sicily 900–500 BC*. Oslo: Unipub.

Armstrong Oma, Kristin. 2010. "Between Trust and Domination: Social Contracts between Humans and Animals." *World Archaeology* 42(2):175–187.

Armstrong Oma, Kristin. forthcoming. "Making Space from the Position of Duty of Care. Early Bronze Age Human—Sheep Entanglements in Norway." In *Multispecies Archaeology*, edited by Suzanne Pilaar Birch. London: Routledge.

Asafu-Adjaye, John, Linus Blomqvist, Stewart Brand, Barry Brook, Ruth DeFries, Erle Ellis, Christopher Foreman, David Keith, Martin Lewis, Mark Lynas, Ted Nordhaus, Roger A. Pielke, Jr., Rachel Pritzker, Joyashree Roy, Mark Sagoff, Michael Shellenberger, Robert Stone, and Peter Teague. 2015. "The Ecomodernist Manifesto." Last modified April 15, 2015. http://www.ecomodernism.org/s/An-Ecomodernist-Manifesto.pdf.

Au, W. W. L., and M. Green. 2000. "Acoustic Interaction of Humpback Whales and Whale-Watching Boats." *Marine Environmental Research* 49:469–481.

Babisch, Wolfgang, Bernd Beule, Marianne Schust, Norbert Kersten, and Hartmut Ising. 2005. "Traffic noise and risk of myocardial infarction." *Epidemiology* 16:33–40.

Bachelard, Gaston. 1938. *La psychanalyse du feu*. Paris: Gallimard.

Bachelard, Gaston. 1942. *L'Eau et les rêves. Essai sur l'imagination de la matière.* Paris, Librairie José Corti.

Bachelard, Gaston. 1943. *L'Air et les songes. Essai sur l'imagination du mouvement.* Paris, Librairie José Corti.

Bagg, Robert, trans. 1978. *The Bakkhai.* Amherst, MA: University of Massachusetts Press.

Bailey, Helen, Bridget Senior, Dave Simmons, Jan Rusin, Gordon Picken, and Paul M. Thompson. 2010. "Assessing underwater noise levels during pile-driving at an offshore windfarm and its potential effects on marine mammals." *Marine Pollution Bulletin.* doi:10.1016/J.marpolbul.2010.01.003.

Bak, Per. 1996. *How Nature Works: The Science of Self-Organized Criticality.* New York: Copernicus.

Baluska, Frantisek, and Stefano Mancuso. 2007. "Plant Neurobiology as a Paradigm Shift Not Only in the Plant Sciences." *Plant Signaling & Behavior* 2(4):205–207.

Barbaras, Renaud. 2008. "Life, Movement, and Desire." *Research in Phenomenology* 38:3–17.

Barbaras, Renaud. 2010. "Life and Exteriority: The Problem of Metabolism." In *Enaction: Toward a New Paradigm for Cognitive Science,* edited by John Stewart, Olivier Gapenne, and Ezequiel A. Di Paolo, 98–122. Cambridge, Mass., London: MIT Press.

Barbaras, Renaud. 2012. "The Phenomenology of Life. Desire as the Being of the Subject." In *The Oxford Handbook of Contemporary Phenomenology,* edited by Dan Zahavi, 94–111. Oxford: Oxford University Press.

Barbaras, Renaud. 2013. "The Subject's Life and the Life of Manifestation. Towards a Privative Biology." *Research in Phenomenology* 43:161–176.

Barber, Jesse R., Kevin R. Crooks, and Kurt M. Fristrup. 2009. "The costs of chronic noise exposure for terrestrial organisms." *TREE* 25(3):180–189.

Barker, Graeme. 2006. *The Agricultural Revolution in Prehistory: Why did Foragers Become Farmers?* Oxford: Oxford University Press.

Barnes, Simon. 2015. "Why Chimps are Almost More Human than We are: As a US Court Grants 'Human Rights' to Chimpanzees, this Account by a Top Naturalist Reveals Why the Animals Are More Like Us than We Think." *Daily Mail,* April 27.

Barney, Erin C., Joel J. Mintzes, and Chiung-Fen Yen. 2005. "Assessing Knowledge, Attitudes, and Behavior Toward Charismatic Megafauna: The Case of Dolphins." *The Journal of Environmental Education,* 36(2):41–55.

Basalla, George. 1988. *The Evolution of Technology.* Cambridge: Cambridge University Press.

Bast, Elizabeth. 2008. "A Climate of War." *Oil Change International,* March 2008.

Bataille, Georges. 1989. *Theory of Religion.* New York: Zone Books.

Bateson, Gregory. 1972. *Steps to an Ecology of Mind.* Northvale, NJ: Jason Aronson.

Beckett, Samuel. 2006. *Waiting for Godot.* New York: Faber and Faber.

Bee, Mark A., and Eli M. Swanson. 2007. "Auditory Masking of Anuran Advertisement Calls by Road Traffic Noise." *Anim. Behav.* 74:1765–1776.

Bekoff, Marc. 2010. *The Animal Manifesto.* California: New World Library.

Bekoff, Marc. 2011. "Animal Minds and the Foible of Human Exceptionalism." *Huffington Post,* August 5.

Berg, Karl, Soraya Delgado, Kathryn A. Cortopassi, Steven R. Beissinger, and Jack W. Bradbury. 2011. "Vertical Transmission of Learned Signatures in a Wild Parrot." *Proc. R. Soc. B* 0932:1471-2954.

Berger, John. 2001. *Why Look at Animals?* London: Penguin.

Bergson, Henri. 1959. *Matter and Memory.* New York: Doubleday.

Berry, Robert J., ed. 2006. *Environmental Stewardship: Critical Perspectives Past and Present.* London: T&T Clarke.

Best, Steven. 2011. "Minding the Animals: Ethology and the Obsolescence of Left Humanism." https://drstevebest.wordpress.com/2011/07/28/minding-the-animals-ethology-and-the-obsolescence-of-left-humanism-2/.

Bird-David, Nurit. 1992. "Beyond "The Original Affluent Society": A Culturalist Reformulation." *Current Anthropology* 33(1):25–47.

Block, Ben. 2015. "Environmentalists spar over corporate ties." *Worldwatch Institute.* www.worldwatch.org

Bonnet, Roger-Maurice, and Lodewijk Woltjer. 2008. *Surviving 1,000 Centuries: Can We Do It?* Berlin, Heidelberg, New York, and Chichester: Springer/Praxis Publishing.

Bowlby, John. 1969. *Attachment and Loss.* Vol.1: *Attachment.* New York: Basic Books.

Boysen, Sally T., and Gary G. Berntson. 1989. "Numerical Competence in a Chimpanzee (*Pan Troglodytes*)." *Journal of Comparative Psychology* 103:23–31.

Bozzetto, Bruno. 1976. *Allegro non troppo.* Film by Bruno Bozzetto.

Bradshaw, John. 2011. *In Defense of Dogs.* London: Allen Lane.

Bratman, Gregory N., J. Paul Hamilton, and Gretchen C. Daily. 2012. "The Impacts of Nature Experience on Human Cognitive Function and Mental Health." *Ann. N.Y. Acad. Sci.* 1249:118–136. doi:10.1111/j.1749-6632.2011.06400.x.

Brenowitz, Eliot A. 1982. "The active space of red-winged blackbird song." *J. Comp Physiol* 147:511–522.

Brentari, Carlo. 2015. *Jakob von Uexküll: The Discovery of the Umwelt Between Biosemiotics and Theoretical Biology.* Dordrecht: Springer.

Brigandt, Ingo. 2005. "The Instinct Concept of the Early Konrad Lorenz." *Journal of the History of Biology* 38:571–608.

Brouček, J. 2014. "Effect of Noise on Performance, Stress, and Behaviour of Animals." *Slovak J. Anim. Sci.* 47:111–123.

Brown, Timothy J., and Paul Handford. 1996. "Acoustical Signal Amplitude Patterns: A Computer Simulation Investigation of the Acoustic Adaptation Hypothesis." *The Condor* 98:608–623.

Bruce, Robin W. 2014. "A Reflection on Biological Thought: Whatever Happened to the Organism?" *Biological Journal of the Linnean Society* 112:354–365.

Brumm, Henrik. 2010. "Anthropogenic Noise: Implications for Conservation." *Encyclopedia of Animal Behavior* 1: pp. 89–93.

Brumm, Henrik, and Dietmar Todt. 2003. "Facing the Rival: Directional Singing Behaviour in Nightingales." *Behaviour* 140:43–53.

Brumm, Henrik, and Hans Slabbekoorn. 2005. Acoustic Communication in Noise. In *Advances in the Study of Behavior*, edited by Peter Slater, Charles Snowdon, Timothy Roper, H. Jane Brockmann, Marc Naguib, vol. 35:151–209. San Diego, Ca.: Academic Press.

Brumm, Henrik, and Peter J. B. Slater. 2006. "Ambient Noise, Motor Fatigue, and Serial Redundancy in Chaffinch Song." *Behav. Ecol. Sociobiol.* 60:475–481.

Bruni, Adrianna, Daniel J. Mennill, and Jennifer R. Foote. 2014. "Dawn Chorus Start Time Variation in a Temperate Bird Community: Relationships with Seasonality, Weather, and Ambient Light." *Journal of Ornithology.* doi:10.1007/s10336-014-1071-7.

Budiansky, Steven. 1992. *The Covenant of the Wild: Why Animals Chose Domestication.* New York: William Morrow.

Budiansky, Steven. 1994. "A Special Relationship: The Coevolution of Human Beings and Domesticated Animals." *JAVMA* 204(3):365–368.

Bugnyar, Thomas, Kijne Maartje, and Kurt Kotrschal. 2001. "Food Calling in Ravens: Are Yells Referential Signals?" *Animal Behaviour* 61:949–58. doi:10.1006/anbe.2000.1668.

Bunkley, Jessie P., Christopher J. W. McClure, Nathan J. Kleist, Clinton D. Francis, and Jesse R. Barber. 2015. "Anthropogenic noise alters bat activity levels and echolocation calls." *Global Ecology and Conservation* 3:62–71.

Burms, Arnold, and Herman De Dijn. 2005. *De Rationaliteit en haar grenzen. Kritiek en Deconstructie.* Leuven: Universitaire Pers Leuven.

Butterfield, Nicholas J. 2007. "Macroevolution and Macroecology through Deep Time." *Palaeontology* 50(1):41–55.

Butterfield, Nicholas J. 2011. "Animals and the Invention of the Phanerozoic Earth System." *Trends in Ecology & Evolution* 26:81–7.

Byrne, Richard W., and Andrew Whiten, eds. 1988. *Machiavellian Intelligence. Social Expertise and the Evolution of Intellect in Monkeys, Apes, and Humans.* (2nd ed. CUP, 1997). Oxford: Clarendon Press.

Caillois, Roger. 1938. *Le mythe et l'homme*. Paris: Gallimard.

Calarco, Matthew. 2008. *Zoographies: The Question of the Animal from Heidegger to Derrida*. New York: Columbia University Press.

Call, Josep, and Michael Tomasello. 2007. *The Gestural Communication of Apes and Monkeys*. New York: Lawrence Erlbaum; Cambridge, MA.: Harvard University Press.

Camaioni, Luigia, Paola Perucchini, Francesca Bellagamba, and Cristina Colonnesi. 2004. "The Role of Declarative Pointing in Developing a Theory of Mind." *Infancy* 5:291–308.

Candea, Matei. 2013. "Habituating Meerkats and Redescribing Animal Behaviour Science." *Theory, Culture & Society* 30(7-8):105–128.

Canguilhem, Georges. 2008. *Knowledge of Life*. New York: Fordham University Press.

Čapek, Karel. 2004. *R.U.R. (Rossum's Universal Robots)*. London: Penguin Classics.

Caro, Tim, and Sheila Girling. 2005. *Antipredator Defenses in Birds and Mammals*. Chicago, Illinois: University of Chicago Press.

Caro, Tim, Jack Darwin, Tavis Forrester, Cynthia Ledoux-Bloom, and Caitlin Wells. 2011. "Conservation in the Anthropocene." *Conservation Biology*. doi:10.1111/j.1523-1739.2011.01752.x.

Carson, Rachel. 1962. *Silent Spring*. New York: Houghton Mifflin Company.

Carver, Steve. 2014. "Making Real Space for Nature: A Continuum Approach to UK Conservation." *ECOS: A Review of Conservation* 35(3/4):4–14. Accessed July 1, 2015. https://www.banc.org.uk/wordpress/ecos-35-34-making-real-space-for-nature-a-continuum-approach-to-uk-conservation-steve-carver/.

Casey, Edward S. 1984. "Habitual Body and Memory in Merleau-Ponty." *Man and World* 17(3):279–97.

Cassidy, Rebecca. 2002. *The Sport of Kings: Kinship, Class and Thoroughbred Breeding in Newmarket*. Cambridge: Cambridge University Press.

Cassidy, Rebecca. 2007. "Introduction: Domestication Reconsidered." In *Where the Wild Things are Now. Domestication Reconsidered,* edited by Rebecca Cassidy, and Molly H. Mullin, 1–25. Oxford & New York: Berg.

Ceballos, Gerardo, Paul E. Ehrlich, Anthony D. Barnosky, Andrés García, Robert M. Pringle, and Todd M. Palmer. 2015. "Accelerated Modern Human-Induced Species Losses: Entering the Sixth Mass Extinction." *Scientific Advances* 1(5), e1400253. doi:10.1126/sciadv.1400253.

Cervantes, Miguel de. 1605 (part 1), 1615 (part 2). *El Ingenioso Hidalgo Don Quixote de la Mancha*. Francisco de Roble Publishers. (In English: The Ingenious Gentleman Don Quixote de la Mancha by Miguel de Cervantes (Saavedra). (Part 1: 1612; Part 2 1620).

Cessford, Gordon R. 2000. "Noise Impact Issues on the Great Walks of New Zealand." USDA Forest Service Proceedings RMRS-P-15-Vol-4:69–76.

Chakraborty, Mukta, Solveig Walløe, Signe Nedergaard, Emma E. Fridel, Torben Dabelsteen, Bente Pakkenberg, Mads F. Bertelsen, Gerry M. Dorrestein, Steven E. Brauth, Sarah E. Durand, and Erich D. Jarvis. 2015. "Core and Shell Song Systems Unique to the Parrot Brain." *PloS ONE* 10(6). doi:10.1371/journal.pone.e0118496.

Chamovitz, Daniel. 2012. *What a Plant Knows: A Field Guide to the Senses*. New York: Scientific American/Farrar, Straus and Giroux.

Chan, Alvin Aaden Yim-Hol, Paulina Giraldo-Perez, Sonja Smith, and Daniel T. Blumstein. 2010. "Anthropogenic Noise Affects Risk Assessment and Attention: The Distracted Prey Hypothesis." *Biology Letters* 6:458–461.

Chapouthier, Georges, and Jean-Claude Nouët, eds. 1998. *The Universal Declaration of Animal Rights: Comments and Intentions*. Paris: Ligue Française des Droits de l'Animal.

Chapron, Guillaume, Petra Kaczensky, John D. C. Linnell, et al. 2014. "Recovery of Large Carnivores in Europe's Modern Human-Dominated Landscapes." *Science* 346(6216):1517–9. Accessed July 13, 2015. doi:1 0.1126/science.1257553.

Chawla, Louise. 1999. "Life Paths into Effective Environmental Action." *The Journal of Environmental Education* 31 (1):15–26. doi:10.1080/00958969909598628.

Cheney, Dorothy L., and Robert M. Seyfarth. 1985. "Vervet Monkey Alarm Calls: Manipulation through Shared Information?" *Behaviour* 93:150–166.

Cheney, Dorothy L., and Robert M. Seyfarth. 2005. "Constraints and Preadaptations in the Earliest Stages of Language Evolution." *Linguistic Review* 22:135–159.

Chew, Sek Jin, David S. Vicario, and Fernando Nottebohm. 1996. "A Large-Capacity Memory System that Recognizes the Calls and Songs of Individual Birds." *Proc. Natl. Acad. Sci. USA* 93:1950–55.

Chrulew, Matthew. 2011. "Managing Love and Death at the Zoo: The Biopolitics of Endangered Species Preservation." *Australian Humanities Review* 50:137–157.

Churchill, Scott. 2005. "Experiencing the Other within the We: Phenomenology with a Bonobo." *Phenomenology* 4:147–170.

CITES. The CITES Appendices, http://www.cites.org/eng/app/index.php. Accessed May 29, 2015.

Clark, Nigel. 2011. *Inhuman Nature: Sociable Life on a Dynamic Planet*. London: Sage.

Clarke, Ellen. 2010. "The Problem of Biological Individuality." *Biological Theory* 5(4):312–25.

Clayton, Nicola S., and Anthony Dickinson. 1998. "Episodic-Like Memory During Cache Recovery by Scrub Jays." *Nature* 395(6699):272–274.

Clayton, Nicola S., and John R. Krebs.1994. "Memory for Spatial and Object-Specific Cues in Food Storing and Non-Storing Birds." *Journal of Comparative Physiology A* 174(3):371–379.

Clingerman, Forrest. 2004. "Beyond the Flowers and the Stones: 'Emplacement' and the Modeling of Nature." *Philosophy in the Contemporary World* 11(2):17–24.

Coetzee, John Maxwell. 1999. *The Lives of Animals*. Princeton, N.J. Chichester: Princeton University Press.

Cohen, Joel E. 2003. "Human Population: The Next Half Century." *Science* 302:1172–1175.

Cohen, Michael P. 2004. "Blues in the Green: Ecocriticism Under Critique." *Environmental History* 9(1):9–36.

Colebrook, Claire. 2014. *Death of the PostHuman: Essays on Extinction*, Vol. 1. Ann Arbor: Open Humanities Press.

Coleman, Alison, and Toby Aykroyd, eds. 2009. *Conference Proceedings: Wild Europe and Large Natural Habitat Areas.* Prague: Wild Europe, European Commission/ EU2009.Cz. Accessed July 13, 2015. http://ec.europa.eu/environment/nature/natura2000/wilderness/pdf/proceedings_wildlife.pdf.

Conniff, Richard. 2014. "Wolves and Bears Stage Comeback in Crowded, Urban Europe." *National Geographic,* December 19. Accessed July 13 2015. http://news.nationalgeographic.com/news/2014/12/141218-/rewilding-europe-wild-animals-science/.

Conway Morris, Simon. 1998. *The Crucible of Creation: The Burgess Shale and the Rise of Animals.* Oxford: Oxford University Press.

Copley, Michael. 2013. "Wind Turbines in Canada Less Lethal to Birds than Other Industrial Sectors." *SNL Energy Power Daily*. October.

Creel, Scott, Jennifer E. Fox, Amanda Hardy, Jennifer Sands, Bob Garrot, and Rolf O. Peterson. 2002. "Snowmobile Activity and Glucocorticoid Stress Responses in Wolves and Elk." *Conservation Biol* 16:809–814.

Crew, Bec. 2015. "Spanish Researchers Are Developing Bladeless Wind Turbines." 15 May 2015. *ScienceAlert*. http://www.sciencealert.com/spanish-researchers-are-developing-bladeless-wind-turbines.

Crutzen, Paul J. 2002. "Geology of Mankind." *Nature* 415, 23. doi:10.1038/415023a.

Crutzen, Paul, and Eugene Stoermer. 2000. "The Anthropocene, Global Change." *IGBP Newsletter* 41:17–18.

Crutzen, Paul J., and Will Steffen 2003. "How Long Have We been in the Anthropocene Era? An Editorial Comment." *Climatic Change* 61:251–257.

Csányi, Vilmos. 2005. *If Dogs Could Talk: Exploring the Canine Mind*, trans. R. E. Quandt. New York: North Point Press.

Cynx Jeffrey, Rebekah Lewis, Benjamin Tavel, and Hanson Tse. 1998. "Amplitude regulation of vocalizations in noise by a songbird, *Taeniopigya guttata*." *Animal Behaviour* 56:107–113.

Dalley, Stephanie, ed. and trans. 1989. *The Epic of Gilgamesh*. In *Myths from Mesopotamia*, edited and translated by Stephanie Dalley. Oxford: Oxford University Press.

Darwin, Charles. 1871/1896. *The Descent of Man*. New York: Appleton.

Darwin, Charles. 1872. *The Expression of the Emotions in Man and Animals*. London: John Murray.

Darwin, Charles. 1890. *The Origin of Species by Means of Natural Selection*. 6th ed. London: John Murray, London.

Darwin, Charles. 1981. *The Descent of Man and Selection in Relation to Sex*. Princeton: Princeton University Press.

Davies, Ben, Christina Pita, David Lusseau, and Colin Hunter. 2010. *The Value of Tourism Expenditure related to the East of Scotland Bottlenose Dolphin Population: Final Report*. Aberdeen: Aberdeen Centre for Environmental Sustainability.

Davis, Susan G. 2006. "Touch the Magic." In *Uncommon Ground: Rethinking the Human Place in Nature*, edited by William Cronon, 204–217. New York: W.W. Norton and Company.

De Sousa, Ronald. 2005. "Biological Individuality." *Croatian Journal of Philosophy* 14:195–218.

de Waal, Frans. 2001. *The Ape and the Sushi Master: Cultural Reflections of a Primatologist*. New York: Basic Books.

de Waal, Frans. 2005. *Our Inner Ape. A Leading Primatologist Explains Why We Are Who We Are*. New York: Riverhead.

de Waal, Frans. 2013. "The Brains of the Animal Kingdom." *The Wall Street Journal*. Accessed March 22. http://www.wsj.com/articles//SB10001424127887323869604578370574285382756.

Deacon, Terrence W. 2011a. *Incomplete Nature. How Mind Emerged From Matter*. New York: Norton & Company.

Deacon, Terrence W. 2011b. "The Importance of What's Missing." *The New Scientist*, November 26, 123–142.

Deinet, Stefanie, Cristina Ieronymidou, Louise McRae, Ian J. Burfield, Ruud P. Foppen, Ben Collen, and Monika Böhm. 2013. *Wildlife comeback in Europe: The recovery of selected mammal and bird species. Final report to Rewilding Europe by ZSL, BirdLife International and the European Bird Census Council*. London, UK: ZSL. Accessed July 13, 2015. http://static.zsl.org/files/wildlife-comeback-in-europe-the-recovery-of-selected-mammal-and-bird-species-2576.pdf.

DeLanda, Manuel. 1992. "Nonorganic Life." In *Zone 6: Incorporations*, edited by Jonathan Crary, and Sanford Kwinter, 129–167. New York: Urzone.

DeLanda, Manuel. 2002. *Intensive Science and Virtual Philosophy*. London: Continuum.

Deleuze, Gilles, and Félix Guattari. 1988. *A Thousand Plateaus: Capitalism and Schizophrenia*, translated by Brian Massumi. London: Athlone Press.

Deliège, Glenn. 2010. "The Cinquefoil Controversy." *Environmental Ethics* 32(1):17–32. doi:10.5840/enviroethics20103213.

Deliège, Glenn. forthcoming. "Contact! Contact! Nature Preservation as the preservation of meaning." *Environmental Values*.

Denham, Tim, and Peter White. 2007. *The Emergence of Agriculture. A Global View*. London and New York: Routledge.

Dennis, Carina. 2002. "A Forage in the Junkyard." *Nature* 420:458–459.

Denton, Michal J., Govindasamy Kumaramanickavel, and Michael Legge. 2013. "Cells As Irreducible Wholes: The Failure of Mechanism and the Possibility of an Organicist Revival." *Biology and Philosophy* 28:31–52.

Derrida, Jacques. 1992. Acts of Literature. New York: Routledge.

Derrida, Jacques. 1995. Points . . . : Interviews, 1974–1994. Standford, Ca: Standford University Press.

Derrida, Jacques. 2006. *L'animal que donc je suis*. Paris: Galilée.

Derrida, Jacques. 2008. *La bête et le souverain, vol 1*. Paris: Galilée.

Derrida, Jacques. 2008. *The Animal That Therefore I Am*. Edited by Marie-Louise Mallet, translated by David Wills. New York: Fordham University Press.

Derrida, Jacques. 2009. *The Beast and the Sovereign, Volume I*. Chicago: Chicago University Press.

Derrida, Jacques. 2010. *La bête et le souverain, vol 2*. Paris: Galilée.

Derrida, Jacques. 2011. *The Beast and the Sovereign, Volume II*. Chicago: Chicago University Press.

Descartes, René. 1970. *Philosophical Letters*. Minneapolis: University of Minnesota Press.

Descartes, René. 1985. *Philosophical Writings*. Vol. 1. Cambridge: Cambridge University Press.

Descartes, René. 1983 (1644). Principles of Philosophy. Dordrecht: Reidel.

Descartes, René. 1991 (1646). "Letter to (the Mrquess of Newcastle) November 23, 1646." In Descartes: Philosophical Letters. Alderly: Clarendon Press.

Descola, Philippe. 2013. *Beyond Nature and Culture*. Translated by Janet Lloyd. Chicago, IL: University of Chicago Press.

Despret, Vinciane. 2004. "The Body We Care For: Figures of Anthropo-Zoo-Genesis." *Body and Society* 10:111–134.

Despret, Vinciane. 2008. "The Becoming of Subjectivity in Animal Worlds." *Subjectivity* 23:123–39.

Diamond, Solomon. 1971. "Gestation of the Instinct Concept." *Journal of the History of Behavioral Sciences* 7(4):323–336.

Díaz, Mario, Antonio Parra, and Clemente Gallardo. 2011. "Serins Respond to Anthropogenic Noise by Increasing Vocal Activity." *Behav. Ecol.* 22:332–336.

Dirzo, Rodolfo, Hilary S. Young, Mauro Galetti, Gerardo Ceballos , Nick J.B Isaac, and Ben Collen. 2014. "Defaunation in the Anthropocene." *Science* 345(6195):401–6. doi:10.1126/science.1251817.

Discovery Institute. "Center on Human Exceptionalism." Accessed in 2015. http://www.discovery.org/human/.

Dodds, Eric R., ed. 1960. *Euripides Bacchae*. Oxford: Oxford University Press.

Dodds, Eric R., ed. 1989. *The Bacchae of Euripides*. Oxford, UK: Clarendon Press, orig. 1994.

Donaldson, Sue, and Will Kymlicka. 2011. *Zoopolis: A Political Theory of Animal Rights*. Oxford: Oxford University Press.

Dransart, Penny Z. 2002. *Earth, Water, Fleece, and Fabric. An Ethnography and Archaeology of Andean Camelid Herding*. New York: Routledge.

Drenthen, Martin. 1999. "The Paradox of Environmental Ethics. Nietzsche's View of Nature and the Wild." *Environmental Ethics* 21(2):163–75.

Drenthen, Martin. 2009a. "Ecological Restoration and Place Attachment: Emplacing Non-Places?" *Environmental Values* 18(3):285–312.

Drenthen, Martin. 2009b. "Fatal Attraction. Wildness in Contemporary Film?" *Environmental Ethics* 31(3):297–315.

Drenthen, Martin. 2011. "Reading Ourselves Through the Land: Landscape Hermeneutics and Ethics of Place." In *Placing Nature on the Borders of Religion, Philosophy, and Ethics*, edited by Forrest Clingerman, and Mark Dixon, 123–38. Farnham: Ashgate Publishing Ltd.

Drenthen, Martin. 2013. "New Nature Narratives." In: *Interpreting Nature. The Emerging Field of Environmental Hermeneutics*, edited by Forrest Clingerman, Martin

Drenthen, Brian Treanor, and David Utsler, 225–41. New York: Fordham University Press.

Drenthen, Martin. 2014. "The Wolf Is Coming! Emplacing a Predator That Is Not (Yet) There." In *Old World and New World Perspective in Environmental Philosophy*, edited by Martin Drenthen, and Jozef Keulartz, 153–174. Dordrecht: Springer.

Drenthen, Martin, and Glenn Deliège. 2014. "Nature Restoration: Avoiding Technological Fixes, Dealing with Moral Conflicts." *Ethical Perspectives* 21(1):101–32.

Drenthen, Martin, Dick Kees, Leo Linnartz, Erwin van Maanen, and Michiel van der Weide. 2015. *De wolf terug. Eng of enerverend?* Utrecht: Kosmos 2015.

Duarte, M. H. L., R. S. Sousa-Lima, R. J. Young, A. Farina, M. Vasconcelos, M. Rodrigues, and N. Pieretti. 2015 "The Impact of Anthropogenic Noise from Open Cast Mining on Atlantic Forest Biophony." *Biological Conservation* 191:623–631.

Dufourcq, Annabelle. 2014. "The Ontological Imaginary: Dehiscence, Sorcery, and Creativity in Merleau-Ponty's Philosophy." *Filozofia* 69:708–718.

Dupré, John, and Maureen A. O'Malley. 2009. "Varieties of Living Things: Life at the Intersection of Lineage and Metabolism." *Philosophy and Theory in Biology* 1(e003):1–25.

Dupré, John. 2007. *The Constituents of Life*. Amsterdam: Van Gorcum.

Dyer, Fred C. 2002. "The Biology of the Dance Language." *Annual Review of Entomology* 47(1):917–49.

Eaton, Heather. 2011. "An Ecological Imaginary: Evolution and Religion in an Ecological Era." Edited by Sigurd Berman and Heather Eaton. *Ecological Awareness, Exploring Religion, Ethica, and Aesthetics*. Studies in Religion and the Environment, vol. 3 Berlin: LIT.

Ehrlich, Paul R. 2009. "Ecoethics: Now Central to All Ethics." *Bioethical Inquiry* 6:417–36.

Eibl-Eibesfeldt, Irenäus. 2007. *Human Ethology*. New York, London: Transaction Publishers.

Einarsson, Niels. 1993. "All animals Are Equal but Some Are Cetaceans: Conservation and Culture Conflict." In *Environmentalism: The View from Anthropology*, edited by Kay Milton, 73–84. Abingdon: Routledge.

Elliot, Larry. 2015. "New Oxfam Report Says Half of Global Wealth Held by the 1%." http://www.theguardian.com/business/2015/jan/19/global-wealth-oxfam-inequality-davos-economic-summit-switzerland. Accessed October 29, 2015.

Elliot, Robert. 1982. "Faking Nature." *Inquiry* 25(1):81–93. doi:10.1080/00201748208601955.

Ellis, Albert Gallatin. 1856. "Advent of the New York Indians into Wisconsin." *Wisconsin Historical Collections.* Volume II:434.

Ellis, Erle C., and Navin Ramankutty. 2008. "Putting People in the Map: Anthropogenic Biomes of the World." Frontiers in Ecology and the Environment 6. doi:10.1890/070062.

Ellis, Erle C. 2011 "Anthropogenic Transformation of the Terrestrial Biosphere." *Philosophical Transactions of the Royal Society A* 369:1010–1035.

Ellis, Erle, and Navin Ramankutty. 2013. "Anthropogenic Biomes." *The Encyclopedia of Earth*. Retrieved from http://www.eoearth.rog/view/article/150128.

Ellul, Jacques. 1964. *The Technological Society*. New York: Vintage.

Erikson, Erik H. 1966. "Ontogeny of Ritualisation in Man." *Philosophical Transactions of the Royal Society of London* 251(B):337–349.

Etter, Paul C. 2012. "Advanced Applications for Underwater Acoustic Modeling." *Advances in Acoustics and Vibration*, article ID 214839. doi:10.1155/2012/214839.

Euripides. 1960. *Euripides: "Bacchae,"* edited with Introduction and Commentary by E. R. Dodds.

Evans, Chris S. 1997. "Referential Signals." *Perspectives in Etholology* 12:99–143.

Evernden, Neil. 1992. *The Social Creation of Nature*. Baltimore and London: The Johns Hopkins University Press.

Fagot, Joël, and Robert Cook. 2006. "Evidence for Large Long-term Memory Capacities in Baboons and Pigeons and Its Implications for Learning and the Evolution of Cognition." *PNAS* 103/46:17564-67.

Farina, Almo 2012. "A Biosemiotic Perspective of the Resource Criterion: Toward a General Theory of Resources." *Biosemiotics* 5:17–32.

Farina, Almo. 2014. *Soundscape Ecology*. Springer, Dordrecht.

Farina, Almo, Silvia Scozzafava, and Brian Napoletano. 2007. "Therapeutic Landscapes: Paradigms and Applications." *J. Mediterranean Ecology* 8:9–13.

Farina, Almo, and Nadia Pieretti. 2014. "Acoustic Codes in Action in a Soundscape Context." *Biosemiotics* 7:321–328.

Farina, Almo, Maria Ceraulo, Christopher Bobryk, Nadia Pieretti, Enza Quinci, and Emanuele Lattanzi. 2015. "Spatial and Temporal Variation of Bird Dawn Chorus and Successive Acoustic Morning Activity in a Mediterranean Landscape." *Bioacoustics* 24(3):269–288.

Farr, Tom G., Cathleen Jones, and Zhen Liu. 2015. "Progress Report: Subsidence in the Central Valley, California." http://water.ca.gov/groundwater/docs/NASA_REPORT.pdf. Accessed October 28, 2015.

Fey, Samuel B., Adam M. Siepielski, Sébastien Nusslé, Kristina Cervantes-Yoshida, Jason L. Hwan, Eric R. Huber, Maxfield J. Fey, Alessandro Catenazzi, and Stephanie M. Carlson. 2015. "Recent Shifts in the Occurrence, Cause, and Magnitude of Animal Mass Mortality Events." *Proceedings of the National Academy of Sciences* 112(4):1083–88.

Fijn, Natasha. 2011. *Living with Herds. Human–Animal Coexistence in Mongolia*. Cambridge: Cambridge University Press.

Flower, Tom, Matthew Gribble, and Amanda Ridley. 2014. "Deception by Flexible Alarm Mimicry in an African Bird." *Science* 344:6183-513–516.

Foreman, Dave. 2004. *Rewilding North America. A Vision for Conservation in the 21st Century*. Washington, Covelo, London: Island Press.

Foreman, Dave. 2013. "Five Feathers for the Cannot Club." In *The Rediscovery of the Wild*, edited by Peter H. Kahn Jr., and Patricia Hasbach, 181–206. Cambridge, London: The MIT Press.

Fossey, Dian. 1983. *Gorillas in the Mist*. Boston: Houghton Mifflin.

Fowler, Karen Joy. 2013. *We Are All Completely Beside Ourselves*. New York: G.P. Putnam's Sons.

Fox, Warwick. 1989. "The Deep Ecology-Ecofeminism Debate and its Parallels." *Environmental Ethics* 11(1):5–25. doi:10.5840/enviroethics198911120.

Francis, CD, Ortega, CP, and Cruz, A. 2009. "Noise Pollution Changes Avian Communities and Species Interactions." *Current Biology* 19:1415–1419.

Franklin, Sarah. 2001. "Sheepwatching." *Anthropology Today* 17(3):3–10.

Franklin, Sarah. 2007. *Dolly Mixtures: The Remaking of Genealogy*. Durham, NC: Duke University Press.

Franklin, Sarah. 2011. "Future Mix: Remodelling Biological Futures." *Humanimalia* 2(2).

Fraser, John, Diana Reiss, Paul Boyle, Katherine Lemcke, Jessica Sickler, Elizabeth Elliott, Barbara Newman, and Sarah Gruber. 2006. "Dolphins in Popular Literature and Media." *Society and Animals* 14(4):321–349.

Frede, Dorothea. 1995. "The Cognitive Role of Phantasia." In *Essays on Aristotle's De Anima*, edited by Martha C. Nussbaum and Amélie Oksenberg Rorty, 279–295. Oxford: Clarendon Press.

Freud, Sigmund. 1958. *The Standard Edition of the Complete Psychological Works of Sigmund Freud*. Vol 13: *Totem and Taboo and Other Works*. London: Hogarth Press.

Freud, Sigmund. 1961. *The Standard Edition of the Complete Psychological Works of Sigmund Freud*. Vol 21: *The Future of an Illusion; Civilization and its Discontents and Other Works*. London: Hogarth Press.

Friese, Carrie. 2013. *Cloning Wild Life: Zoos, Captivity, and the Future of Endangered Animals*. New York: NYU Press.

Frisk, G. V. 2012. "Noiseonomics: The Relationship between Ambient Noise Levels in the Sea and Global Economic Trends." *Nature*. doi:10.1038/srep00437.

Gaita, Raimond. 2002. *The Philosopher's Dog*. Melbourne: Text Publishing.

Gall, M.D., K. L. Ronald, E. S. Bestrom, and J. R. Lucas. 2012. "Effects of Habitat and Urbanization on the Active Space of Brown-Headed Cowbird Song." *J. Acoust. Soc. Am.* 132(6):4053–4062.

Gardner, R. Allen, and Beatrix T. Gardner. 1971. "Two-Way Communication with an Infant Chimpanzee." In *Behavior of Nonhuman Primates: Modern Research Trends*, edited by Allan M. Schrier, and Fred Stollnitz. Vol. 4, 117–183. New York: Academic Press.

Gardner, R. Allen, Beatrix T. Gardner, and Thomas E. Van Cantfort, eds. 1989. *Teaching Sign Language to Chimpanzees*. Albany: State University of New York Press.

Gare, Arran. 2008. "Approaches to the Question 'What is Life?': Reconciling Theoretical Biology with Philosophical Biology." *Cosmos and History: The Journal of Natural and Social Philosophy* 4(1–2):53–77.

Gehlen, Arnold. 1988. *Man. His Nature and Place in the World*. New York: Columbia University Press.

Gentner, Timothy Q., Kimberly M. Fenn, Daniel Margoliash, and Howard C. Nusbaum. 2006. "Recursive Syntactic Pattern Learning by Songbirds." *Nature*, 440 (7088), 1204–1207.

Gill, Sharon A., and Andrea M-K. Bierema. 2013. "On the Meaning of Alarm Calls: A Review of Functional Reference in Avian Alarm Calling." *Ethology* 119(6):449–461. doi:10.1111/eth. 12097.

Gillespie-Lynch, Kristen, Patricia M. Greenfield, Heidi Lyn, and Sue Savage-Rumbaugh. 2014. "Gestural and Symbolic Development among Apes and Humans: Support for a Multimodal Theory of Language Evolution." *Frontiers in Psychology* 5:1228. doi:10.3389/fpsyg.2014.01228.

Glendinning, Simon. 1998. *On Being With Others: Heidegger, Derrida, Wittgenstein*. New York: Routledge.

Goines, L., and L. Hagler. 2007. "Noise Pollution: A Modern Plague." *Southern Medical Journal* 100:287–294.

Goldstein, Kurt. 1995. *The Organism. A Holistic Approach to Biology Derived from Pathological Data in Man*, with a foreword by Oliver Sacks. New York: Zone Books.

Goodall, Jane. 1986. *The Chimpanzees of Gombe: Patterns of Behaviour*. Cambridge, MA: Harvard University Press.

Goodall, Jane. *Guardian*, 12 June 2015. http://www.theguardian.com/world/2015/jun/12/jane-goodall-us-chimpanzees-endangered.

Goosem, M. 2012. "Mitigating the Impacts of Rainforest Roads in Queensland's Wet Tropic: Effective or Are Further Evaluations and New Mitigation Strategies Required?" *Ecological Management & Restoration* 13:254–258.

Gottieb, Roger. 2004. *This Sacred Earth: Religion, Nature, Environment*. New York: Psychology Press.

Green, D. M., and J. A. Swets. 1966. *Signal Detection Theory and Psychophysics*. New York: Wiley

Grene, Marjorie. 1965. "Beyond Darwinism: Portmann's Thought." *Commentary* 40, November 1.

*Grizzly Man*. 2005. Directed by Werner Herzog. Hollywood, California. Lionsgate Entertainment. DVD.

Gudeman, Steven. 1986. *Economics as Culture: Models and Metaphors of Livelihood*. London: Routledge and Kegan Paul.

Haff, Peter K. 2013 "Technology As a Geological Phenomenon: Implications for Human Well-Being." *Geological Society, London, Special Publications* 395.

Haldane, John Burdon Sanderson. 1953. "Rituel Humain et Communication Animale." *Diogene* 4:77–93

Halliday, M. A. K. 2001. "New Ways of Meaning: The Challenge to Applied Linguistics." In *The Ecolinguistics Reader*. Edited by Alwin Fill and Peter Mulhausler. London: Continuum, 175–202.

Hameed, Bassim H., L. F. Lai, and L. H. Chin. 2009. "Production of Biodiesel from Palm Oil *Elaeis guineensis* using heterogeneous catalyst: an optimized process." *Fuel Processing Technology* 90(4):606–10.

Hanna, Robert, and Evan Thompson. 2003. "The Mind-Body-Body Problem." *Theoria et Historia Scientiarum: International Journal for Interdisciplinary Studies* 7:23–42.

Haraway, Donna J. 1978. "Animal Sociology and a Natural Economy of the Body Politic, Part II: The Past Is the Contested Zone: Human Nature and Theories of Production and Reproduction in Primate Behavior Studies." *Signs: Journal of Women in Culture and Society* 4(1):37–60.

Haraway, Donna J. 1989. *Primate Visions: Gender, Race, and Nature in the World of Modern Science*. New York: Routledge.

Haraway, Donna J. 1992. *Primate Visions: Gender, Race, and Nature in the Word of Modern Science*. London, New York: Verso.

Haraway, Donna J. 2003. *The Companion Species Manifesto: Dogs, People, and Significant Otherness*. Chicago: Prickly Paradigm Press.

Haraway, Donna J. 2008. *When Species Meet*. Minneapolis, MN: University of Minnesota Press.

Haraway, Donna J. 2015. "Anthropocene, Capitalocene, Plantationocene, Chthulucene: Making Kin." *Environmental Humanities* 6:159–165.

Hardoon, Deborah. 2015. "Wealth: Having it All and Wanting More." http://policy-practice.oxfam.org.uk/publications/wealth-having-it-all-and-wanting-more-338125. Accessed October 27, 2015.

Hare, Brian, and Michael Tomasello. 2006. "Behavioural Genetics of Dog Cognition: Human-Like Social Skills in Dogs Are Heritable and Derived." In *The Dog and its Genome*, edited by Elaine A. Ostrander, Urs Giger, and Kerstin Lindblad-Toh, 497–514. Cold Spring Harbor, New York: Coldspring Harbor Laboratory Press.

Hare, Brian, and Vanessa Woods. 2013. *The Genius of Dogs: How Dogs Are Smarter than You Think*. New York: Penguin.

Hartig, Terry, Richard Mitchell, Sjerp de Vries, and Howard Frumkin. 2014. "Nature and Health." *Annual Review of Public Health* 35:207–228. doi:10.1146/annurev-publhealth-032013-182443.

Hartmann, Nicolai. 1950. *Philosophie der Natur: Abriss der speziellen Kategorienlehre*. Berlin: de Gruyter.

Hatley, James. 2004. "The Uncanny Goodness of Being Edible to Bears." In *Rethinking Nature: Essays in Environmental Philosophy*, edited by Bruce Foltz, and Robert Frodeman, 13–31. Indianapolis: Indiana University Press.

Hearne, Vicki. 1987. *Adam's Task: Calling Animals by Name*. New York: Knopf.

Hearne, Vicki. 2000. *Adam's Task: Calling Animals by Name*. Pleasantville, NY: Akadine Press.

Hearne, Vicki. 2007. *Adam's Task: Calling Animals by Name*. New York: Skyhorse Publishing.

Heatherington, Tracey 2008. "Cloning the Wild Mouflon." *Anthropology Today*, 24(1):9–14.

Hedblom, Marcus, Erik Heyman, Antonsson, and Bengt Gunnarsson. 2014. "Bird Song Diversity Influences Young People's Appreciation of Urban Landscapes." *Urban Forestry & Urban Greening* 13:469–474.

Heidegger, Martin. 1962. *Being and Time*, translated by John Macquarrie and Edward Robinson. Malden, MA: Blackwell.

Heidegger, Martin. 1977. "The Question Concerning Technology." In *Basic Writings*, ed. David Farrell Krell, 311–341. New York: Harper & Row.

Heidegger, Martin. 1983. *Die Grundbegriffe der Metaphysik*. (*Gesamtausgabe*, Band 29/30) Frankfurt am Main: Klostermann.

Heidegger, Martin. 1993 [1947]. "Letter on Humanism." In *Basic Writings*, ed. David Farrell Krell. San Francisco: Harper Collins.

Heidegger, Martin. 1995 [1929]. *The Fundamental Concepts of Metaphysics*. Bloomington, IN: Indiana University Press.

Henry, Michel. 2000. *Incarnation*. Paris: Seuil.

Hird, Myra. 2009. *The Origins of Sociable Life: Evolution after Science Studies*. London: Palgrave Macmillan.

Hirschboeck, Paula. Zen Buddhist priest. 2015. Personal communication.

Hobbes, Thomas. 1994. *Leviathan*. Edited by Edwin Curley. Indianapolis, IN: Hackett Publishers.

Hobson, Kersty. 2007. "Political Animals? On Animals As Subjects in an Enlarged Political Geography." *Political Geography* 26–3: 250–67.

Höchtl, Franz, Susanne Lehringer, and Werner Konold. 2005. "'Wilderness': What It Means When It Becomes a Reality—A Case Study from the Southwestern Alps." *Landscape and Urban Planning* 70:85–95.

Hodder, Ian. 1990. *The Domestication of Europe. Structure and Contingency in Neolithic Societies*. Oxford, UK and Cambridge, USA: Blackwell.

Hodos, William, and C. B. G. Campbell. 1969. "Scala Naturae: Why There Is No Theory in Comparative Psychology." *Psychological Review* 76(4):337–350.

Hoffmeyer, Jesper. 1996. *Signs of Meaning in the Universe*. Bloomington, IN: Indiana University Press.

Hoffmeyer, Jesper. 2008. *Biosemiotics: The Signs of Life and the Life of Signs*. Scranton, PA: University of Scranton Press.

Hoffmeyer, Jesper. 2012. "I Am Plural." In *Semiotics in the Wild: Essays in Honour of Kalevi Kull on the Occasion of his Sixtieth Birthday*, 176–182. Tartu: University of Tartu Press.

Hoffmeyer, Jesper. 2015. "Semiotic Scaffolding of Multicellularity." *Biosemiotics*, 8(2):159–71.

Hollén Linda I., and Andrew N. Radford. 2009. "The Development of Alarm Call Behaviour in Mammals and Birds." *Animal Behaviour* 78(4):791–800. doi:10.1016/j.anbehav.2009.07.021.

Holt, M. N., D. P. Noren, V. Veirs, C. K. Emmons, S. Veirs. 2008. "Speaking Up: Killer Whales (*Orcinus orca*) increase their call amplitude in response to vessel noise." *J. Acoust. Soc. Am.* 125(1): EL27–EL32.

Holy See. 2015. "On Care for our Common Home" *Laudato Si.'* Encyclical Letter. Rome: Vatican. http://w2.vatican.va/content/francesco/en/encyclicals/documents/papa-francesco_20150524_enciclica-laudato-si.html.

Hopkins, William D., Jamie L. Russell, and Jennifer A. Schaeffer. 2012. "The Neural and Cognitive Correlates of Aimed Throwing in Chimpanzees: A Magnetic Resonance Image and Behavioural Study on a Unique Form of Social Tool Use." *Phil. Trans. R. Soc. B* 367(1585):37–47.

Howard, C., and E. C. M. Parsons. 2006. "Attitudes of Scottish City Inhabitants to Cetacean Conservation." *Biodiversity and Conservation* 15(14):4335–4356.

Howard, Len. 1952. *Birds as Individuals*. London: Collins Press.

Howard, Len. 1956. *Living with Birds*. London: Collins Press.

Hu, Jane C. 2014. "What Do Talking Apes Really Tell Us?" *Slate*, 20 August. http://www.slate.com/articles/health_and_science/science/2014/08/koko_kanzi_and_ape_language_research_criticism_of_working_conditions_and.html.

Huneman, Philippe, and Charles T. Wolfe. 2010. "The Concept of Organism: Historical, Philosophical, Scientific Perspectives." *History and Philosophy of the Life Sciences* 32(2/3):147–154.

Hunt, Gavin R. 1996. "Manufacture and Use of Hook-Tools by New Caledonian Crows." *Nature* 379(6562):249–251.

Hunziker, M. 1995. "The Spontaneous Reafforestation in Abandoned Agricultural Lands: Perception and Aesthetic Assessment by Locals and Tourists." *Landscape and Urban Planning* 31:399–410.

Husserl, Edmund. 1910. *Logische Untersuchungen. Zweiter Band, Erster Teil: Untersuchungen zur Phänomenologie und Theorie der Erkenntnis,* 1901, Max Niemeyer Verlag, Tübingen, 7 Auflage, 1993.

Husserl, Edmund. 1950. *Ideen zu einer reinen Phänomenologie und phänomenologischen Philosophie, Erstes Buch: Allgemeine Einführung in die reine Phänomenologie.* (*Husserliana* III). The Hague: M. Nijhoff.

Husserl, Edmund. 1954. *Die Krisis der europäischen Wissenschaften und die transzendentale Phänomenologie.* Husserliana vol. 29. Edited by Reinhold N. Smid. Dordrecht: Kluwer.

Husserl, Edmund. 1980. *Phantasie, Bildbewußtsein, Erinnerung.* (Husserliana XXIII). The Hague: M. Nijhoff.

Husserl, Edmund. 1993. *Die Krisis der Europäischen Wissenschaften und die Transzendentale Phänomenologie.* Husserliana vol 29. Edited by Reinhold N. Smid. Dordrecht: Kluwer.

Huxley, Julian. 1966. "A Discussion on Ritualization of Behaviour in Animals and Man." *Philosophical Transactions of the Royal Society of London,* Series B, Biological Sciences 251(772):249–271.

Igic, Branislav, and Robert D. Magrath. 2014. "A Songbird Mimics Different Heterospecific Alarm Calls in Response to Different Types of Threat." *Behavioral Ecology.* doi:10.1093/beheco/aru018.

Ingensiep, Hans Werner. 2001. *Geschichte der Pflanzenseele. Philosophische und biologische Entwürfe von der Antike bis zur Gegenwart.* Kröner: Stuttgart.

Ingold, Tim. 2000. *The Perception of the Environment: Essays in Livelihood, Dwelling, and Skill.* London and New York: Routledge.

Ingold, Tim. 2000. "From Trust to Domination: An Alternative History of Human–Animal Relations." In *The Perception of the Environment. Essays in Livelihood, Dwelling and Skill,* edited by Tim Ingold, 61–76. London & New York: Routledge.

IPCC. 2013. "Summary for Policymakers." In *Climate Change 2013: The Physical Science Basis. Contribution of Working Group I to the Fifth Assessment Report of the Intergovernmental Panel on Climate Change,* edited by T.F. Stocker, D. Qin, G.-K. Plattner, M. Tignor, S.K. Allen, J. Boschung, A. Nauels, Y. Xia, V. Bex, and P.M. Midgley, 1–28. Cambridge, United Kingdom, and New York, NY: Cambridge University Press.

IUCN. "Tursiops truncatus." http://www.iucnredlist.org/details/22563/0. Accessed May 29, 2015.

Iveson, Richard. 2010. "Animals in Looking-Glass World: Fables of Überhumanism and Posthumanism in Heidegger and Nietzsche." *Humanimalia: A Journal of Human/Animal Interface Studies* 2:46–85.

Iveson, Richard. 2012. "On the Importance of Heidegger's Anthropogenesis, and of Moving Beyond It." *Unruly Creatures 2: Creative Revolutions* at the Natural History Museum in London, 18 June.

Izard, Véronique, Pierre Pica, Stanislas Dehaene, Danielle Hinchey, and E. S. Spelke. 2011. "Geometry as a universal mental construction." In *Space, Time and Number in the Brain: Searching for the Foundations of Mathematical Thought,* edited by Elizabeth Brannon and Stanislas Dehaene, 319–332. Boston: Academic Press (Elsevier).

Jackson, Michael. 1983. "Thinking Through the Body: An Essay on Understanding Metaphor." *Social Analysis* 14:127–148.

Jacobson-Tepfer, Esther. 2015. *The Hunter, the Stag, and the Mother of Animals: Image, Monument, and Landscape in Ancient North Asia.* New York: Oxford.

Jarvis, Erich D. 2015. "Core and Shell Song Systems Unique to the Parrot Brain." *PloS ONE* 10(6). doi:10.1371/journal.pone.e0118496.

Jarvis, Erich D., Onur Güntürkün, Laura Bruce, András Csillag, Harvey Karten, Wayne Kuenzel, Loreta Medina, et al. 2005."Avian brains and a new understanding of vertebrate brain evolution." *Nature Reviews Neuroscience* 6(2):151–159. doi:10.1038/nrn1606.

Jensen, Derek. 2013. "Age of the Sociopath." *Earth Island Journal,* spring 2013.

Johannsen, Niels. 2012. "Archaeology and the Inanimate Agency Proposition: A Critique and a Suggestion." In *Excavating the Mind. Cross-sections through Culture, Cognition and Materiality,* edited by Niels Johannsen, Mads Jessen, and Helle Juel Jensen, 305–347. Aarhus: Aarhus University Press.

Johannsen, Thomas K. 2012. *The Powers of Aristotle's Soul.* Oxford: Oxford University Press.

Johnston, Alison, S. C. P. Aonghais, Lucy J. Wright, Elizabeth M. Humphreys, and Niall H. K. Burton. 2014. "Modelling Flight Heights of Marine Birds to More Accurately Assess Collision Risk with Offshore Wind Turbines." *Journal of Applied Ecology* 51(1):31–41. doi:10.1111/1365-2664.

Jonas, Hans. 1966. *The Phenomenon of Life: Toward a Philosophical Biology.* Evanston, Illinois: Northwestern University Press.

Jones, Andy, and Colin Richards. 2003. "Animals into Ancestors: Domestication, Food and Identity in Late Neolithic Orkney." In *Food, Culture and Identity in the Neolithic and Early Bronze Age,* edited by Mike Parker Pearson, 45–52. BAR International Series. Oxford: Archaeopress.

Joo, W., S. H. Gage, and E.P. Kasten. 2011. "Analysis and Interpretation of Variability in Soundscapes along an Urban-Rural Gradient." *Landscape and Urban Planning* 103:259–276.

Kafka, Franz. 1959. "A Report to an Academy." In *The Penal Colony: Stories and Short Pieces,* translated by Willa and Edwin Muir. New York: Schocken Books.

Kant, Immanuel. 2007. *Anthropology, History, and Education.* Edited/translated by Günterzöller and Robert B. Louden. Cambridge, UK: Cambridge University Press.

Kaplan, Gisela. 2001. "New Social Movements." In *Encyclopedia of Modern European Social History* Vol.3:289–300. New York: Charles Scribner's Sons Reference Books.

Kaplan, Gisela. 2009. "Animals and Music: Between Cultural Definitions and Sensory Evidence." *Sign Systems Studies* 37(3/4):75–101.

Kaplan, Gisela. 2013. "Pointing Gesture in a Bird—Merely Instrumental or a Cognitively Complex Behavior?" *Current Zoology* 57(4):453–467.

Kaplan, Gisela 2014. "Animal Communication. Invited Advanced Review." *WIREs Cognitive Science.* doi:10.1002/wcs.1321, 17pp.

Kaplan, Gisela. 2015. *Bird Minds: Cognition and Behaviour of Australian Native Species.* Melbourne, Australia: CSIRO Publishing.

Kaplan, Gisela, and Lesley J. Rogers. 2003. *Gene Worship. Moving Beyond the Nature/Nurture Debate over Genes, Brain and Gender.* New York and London: Other Press LLC.

Kaplan, Gisela, and Lesley J. Rogers. 2004. "Charles Darwin and Animal Behavior." In *Encyclopedia of Animal Behavior* (3 vols), edited by Marc Bekoff and Jane Goodall, vol. 2:471–479. Westport, CT: Greenwood Publishing.

Kaplan, Gisela, and Lesley J. Rogers. 2013. "Stability of Referential Signalling across Time and Locations: Testing Alarm Calls of Australian Magpies (*Gymnorhina tibicen*) in urban and rural Australia and in Fiji." *PeerJ.* 1:e112. doi:10.7717/peerj.112.

Kaplan, Stephen, and Rachel Kaplan. 1989. *The Experience of Nature: A Psychological Perspective.* Cambridge, UK: Cambridge University Press.

Kappala-Ransamy, Gemma. 2011. "Nim Chimpsky: The Chimp They Tried to Turn into a Human." *The Observer,* 23 July. www.theguardian.com/film/2011/jul/24/project-nim-chimpsky-chimpanzee-language.

Katz, Eric. 2000. "The Big Lie: The Human Restoration of Nature." In *Environmental Restoration. Ethics, Theory and Practice,* edited by William Throop, 83–93. New York: Humanity Books.

Keim, Brandon. 2015. "Chimpanzee Rights Get a Day in Court." *Science* 27. http://www.wired.com/2015/05/chimpanzee-rights-get-day-court/WIRED.

Keller, Georg B., and Richard H. R. Hahnloser. 2009. "Neural Processing of Auditory Feedback During Vocal Practice in a Songbird." *Nature* 457:187–190. doi:10.1038/nature07467

Kendon, Adam. 2014. "Semiotic Diversity in Utterance Production and the Concept of 'Language'." *Philosophical Transactions of the Royal Society B: Biological Sciences* 369(1651):20130293.

Kenny, Anthony. 1993. *Aquinas on Mind*. London: Routledge.

King, J. H. Roger. 2003. "Toward an Ethic of the Domesticated Environment." *Philosophy & Geography* 6(1):3–14. doi:10.1080/1090377032000063289.

King, Stefanie, and Vincent Janik. 2013. "Bottlenose Dolphins Can Use Learned Vocal Labels to Address Each Other," *Proceedings of the National Academy of Sciences,* 110–32: 13216–21.

King, Ursula. 2014. "The Zest for Life: A Contemporary Exploration of a Generative Theme in Teilhard's Work." In *From Teilhard to Omega,* edited by Ilia Delio. New York: Orbis Books.

Kirksey, S. Eben, and Stefan Helmreich. 2010. "The Emergence of Multispecies Ethnography." *Cultural Anthropology* 25(4):545–576.

Klein, Naomi. 2014. *This Changes Everything: Capitalism vs. The Climate.* New York: Simon and Schuster.

Kleisner, Karel and Anton Markoš. 2005. "Semetic Rings: Towards the New Concept of Mimetic Resemblances." *Theory in Biosciences* 123:209–222.

Klinck, H., S. L. Nieukirk, D. K. Mellinger, K. Klinck, H. Matsumoto, and R. P. Dziak. 2012. "Seasonal Presence of Cetaceans and Ambient Noise Levels in Polar Waters of the North Atlantic." *J. Acoust. Soc. Am. 132*(3): EL176–EL181.

Klump, G. M. 1996. "Bird Communication in the Noisy World." In *Ecology and Evolution of Acoustic Communication in Birds,* edited by D. E. Kroodsma, and E. H. Miller, 321–338. Ithaca, New York, USA: Cornell University Press.

Kloepper, L. N., and A. M. Simmons. 2014. "Bioacoustic Monitoring Contributes to an Understanding of Climate Change," *Acoustic Today,* Summer 2014: 8–15.

Klyza, Christopher M. 2001. *Wilderness Comes Home.* Hanover, NH: University Press of New England.

Knight, John. 2005. "Introduction." In *Animals in Person: Cultural Perspectives on Human–Animal Intimacy,* edited by John S. Knight, 1–13. Oxford & New York: Berg.

Knight, John. 2006. "Monkey Mountain as a Megazoo: Analyzing the Naturalistic Claims of "Wild Monkey Parks" in Japan." *Society and Animals* 14(3):245–264.

Koenig, Otto. 1975. *Urmotiv Auge*. Zürich, München: Piper.

Köhler, Wolfgang. 1927. *The Mentality of Apes*. New York: Harcourt Brace.

Kolbert, Elizabeth. 2010. The Anthropocene Debate: Marking Humanity's Impact. *Yale Environment 360* (Yale School of Forestry and Environmental Studies). http// e360.yale.edu/feature/the_anthropocene_debate_marking_humanitys_impact/ 2274/.

Kopenawa, Davi, and Bruce Alberti. 2013. *The Falling Sky. Words of a Yanomami Shaman.* Translated by Nicholas Elliott, and Alison Dundy. Cambridge, MA, and London: Harvard University Press.

Koper, R. P., and S. Plön. 2012. "The Potential Impacts of Anthropogenic Noise on Marine Animals and Recommendations for Research in South Africa." *EWT Research & Technical Paper No. 1.* Endangered Wildlife Trust, South Africa.

Korner-Nievergelt, Fränzi, Robert Brinkmann, Ivo Niermann, and Oliver Behr. 2013. "Estimating Bat and Bird Mortality Occurring at Wind Energy Turbines from Covariates and Carcass Searches Using Mixture Models." *PLoS ONE* 8(7):e67997. doi:10.1371/journal.pone.0067997.

Krampen, Martin. 2010. "Phytosemiotics." In *Essential Readings in Biosemiotics. Anthology and Commentary,* ed. by Donald Favareau. Springer: Dordrecht.

Krause, Bernie. 2012. *The Great Animal Orchestra.* New York: Little, Brown and Company.

Krijgsveld, Karen L., Kirsten Akershoek, Femke Schenk, Femke Dijk, and Sjoerd Dirksen. 2009. "Collision Risk of Birds with Modern Large Wind Turbines." *Ardea* 97(3):357–366. doi:http://dx.doi.org/10.5253/078.097.0311.

Kull, Kalevi. 1998. "Semiotic Ecology: Different Natures in the Semiosphere." *Sign Systems Studies* 26:344–371.

Kull, Kalevi. 1999. "Umwelt and Evolution: From Uexküll to Post-Darwinism." In *Semiosis, Evolution, Energy*, edited by Edwina Taborsky, 53–70. Aachen: Shaker Verlag.

Kull, Kalevi. 2000. "An Introduction to Phytosemiotics: Semiotic Botany and Vegetative Sign Systems." *Sign Systems Studies* 28:326–350.

Kull, Kalevi, and Peeter Torop. 2011. "Biotranslation: Translation Between Umwelten." In *Readings in Zoosemiotics*, edited by Timo Maran, Dario Martinelli, and Aleksei Turovski, 411–425. Berlin: De Gruyter Mouton.

Kumar, Satish. 2013. "The Ecozoic Era." *Resurgence and Ecologist* 279(July/August):1.

Lam, Man K., Kok Tat Tan, Keat-Teong Lee, and Abdul R. Mohamed. 2008. "Malaysian Palm oil: Surviving the food versus fuel dispute for a sustainable future (Report)." *Renewable and Sustainable Energy Review.* 13(6-7):1456–1464. doi:10.1016/j.rser.2008.09.009.

Lampe, U., T. Schmoll, A. Frnzke, and K. Reinhold. 2012. "Staying Tuned: Grasshoppers from Noisy Roadside Habitats Produce Courtship Signals with Elevated Frequency Components." *Functional Ecology* 26:1348–1354.

Lampe, U., K. Reinhold, and T. Schmoll. 2014. "How Grasshoppers Respond to Road Noise: Development Plasticity and Population Differentiation in Acoustic Signaling." *Functional Ecology* 28:660–668.

Land, Nick. 2012 "Barker speaks: The CCRU Interview with Professor D. C. Barker." In *Fanged Noumena: Collected Writings 1987- 2007*, 493–505. Falmouth: Urbanomic.

Langerholc, John. 1991. "Facial Mimicry in the Animal Kingdom." *Bolletino di zoologia* 58(3):185–204.

Langmuir, Charles H., and Wally Broecker. 2012. *How to Build a Habitable Planet*, revised edition. Princeton: Princeton University Press.

Latour, Bruno. 2013. *An Inquiry into Modes of Existence: An Anthropology of the Moderns.* Cambridge, MA; London, England: Harvard University Press.

Lazcano, Antonio, Arturo Becerra, and Luis Delaye. 2011. "Alarmones." In *Chimeras and Consciousness: Evolution of the Sensory Self*, edited by Lynn Margulis, Celeste A. Asikainen, and Wolfgang E. Krumbein, 35–43. Cambridge: MA: MIT Press.

Le Bon, Gustave. 1896. *The Crowd: A Study on the Popular Mind*. New York: Macmillan.

Leakey, Richard, and Roger Lewis. 1996. *The Sixth Extinction: Patterns of Life and the Future of Humankind.* London: Weidenfeld and Nicolson.

Leavens, David, Jamie L. Russell, and William D. Hopkins. 2005. "Intentionality as Measured in the Persistence and Elaboration of Communication by Chimpanzees *Pan Troglodytes*." *Child Development* 76:291– 306.

Leder, Drew. 1990. *The Absent Body*. Chicago: The University of Chicago Press.

Lehrer, Miriam, and Raymond Campan. 2004. "Shape Discrimination by Wasps (*Paravespula Germanica*) at the Food Source: Generalizations among Various Types of Contrast." *Journal of Comparative Physiology A* 190(8):651–663. doi:10.1007/s00359-004-0523-6.

Lengagne, T., and P. J. B. Slater. 2002. "The Effects of Rain on Acoustics Communication: Tawny Owls Have Good Reason for Calling Less in Wet Weather." *Proc. Roy. Soc. Lond. B*: Biol. 269:2121–2125.

Lenton, Timothy M., and Andrew J. Watson. 2011. *Revolutions that Made the Earth.* Oxford: Oxford University Press.

Leopold, Aldo. 1986. *A Sand County Almanac*. New York: Ballantine Books.

Leroi-Gourhan, André. 1967. *Treasures of Prehistoric Art.* Translated by Norbert Guterman. New York: Harry N. Abrams.

Leroi-Gourhan, André. 1971 [1943]. *Évolution et Techniques, Vol. 1: L'Homme et la Matière*. Paris: Albin Michel.

Lestel, Dominique. 2002. "The Biosemiotics and Phylogenesis of Culture." *Social Science Information* 41(1):35–68.

Lestel, Dominique. 2007. *L'Animalité*. Paris: Éditions de L'Herne.

Lestel, Dominique. 2011. "What Capabilities for the Animal?" *Biosemiotics* 4(1):83–102.

Lestel, Dominique. 2011. "The Biosemiotics and Pylogenesis of Culture." In *Readings in Zoosemiotics*, edited by Timo Maran, Dario Martinelli, and Aleksei Turovski, 377–409. Berlin: De Gruyter Mouton.

Lestel, Dominique. 2013. "The Withering of Shared Life through the Loss of Biodiversity." *Social Science Information* 52:307–325.

Levinas, Emmanuel. 1930. *La théorie de l'intuition dans la phénoménologie de Husserl*. Paris: Vrin.

Lévi-Strauss, Claude. 1964. *Totemism*. London: Merlin Press.

Lewontin, Richard C. 1991. *The Doctrine of DNA*. London: Penguin Books.

Lewontin, Richard C., Steven Rose, and Leon J. Kamin. 1984. *Not in Our Genes: Biology, Ideology, and Human Nature*. New York: Pantheon.

Lieberman, Philip. 2015. "Language Did Not Spring Forth 100,000 Years Ago." *PLoS Biol.* 13(2): doi:10.1371/journal.pbio.11002064.

Linden, Eugene. 1987. *Silent Partners: The Legacy of the Ape Language Experiments*. New York: Ballentine Books.

Lines-Kelly, Rebecca. 2004. "Soil: Our Common Ground—A Humanities Perspective." SuperSoil 2004: 3rd Australian New Zealand Soils Conference, 5–9 December 2004, University of Sydney, Australia. http://www.regional.org.au/au/asssi/supersoil2004/keynote/lineskelly.htm. Accessed October 30, 2015.

Lingis, Alphonso. 2003. "Animal Body, Inhuman Face." In *Zoontologies: The Question of the Animal*, edited by Cary Wolfe, 165–182. Minneapolis: University of Minnesota Press.

Linzey Andrew, and Paul Barry Clarke, eds. 1990. *Animal Rights: A Historical Anthology*. Chapter 14: Jeremy Bentham, "Duty to Minimize Suffering," 135–137. New York: Columbia University Press.

Liszkowski, Ulf, Malinda Carpenter, Anne Henning, Tricia Striano, and Michael Tomasello. 2004. "Twelve-Month-Olds Point to Share Attention and Interest." *Developmental Science* 7(3):297–307.

Locke, Devin P., LaDeana W. Hillier, Wesley C. Warren, et al. 2011. "Comparative and Demographic Analysis of Orang-utan Genomes." *Nature* 469:529–35. doi:10.1038/nature09687.

Lombard, E. 1911. "Le signe de l'elevation e la voix." *Annales des Maladies de l'Oreille e du Larynx* 37:101–119.

Lorenz, Konrad. 1949. *King Solomon's Ring: New Light on Animal Ways*. London: Routledge.

Lorenz, Konrad. 1981. *The Foundations of Ethology*. London, New York/Wien: Springer.

Lorenz, Konrad. 2002. *On Aggression*. London, New York: Routledge.

Lorimer, Jamie. 2015. *Wildlife in the Anthropocene: Conservation after Nature*. London: University of Minnesota Press.

Lotman, Yuri M. 1990. *Universe of the Mind: A Semiotic Theory of Culture*, translated by Ann Shukman and with an Introduction by Umberto Eco. Bloomington, IN: Indiana University Press.

Louv, Richard. 2005. *Last Child in the Woods: Saving Our Children from Nature-Deficit Disorder*. Chapel Hill: Algonquin Books.

Lovelock, James. 2006. *The Revenge of Gaia: Earth's Climate Crisis and the Fate of Humanity*. New York: Basic Books.

Lupo, C., L. Lodi, G. Paluffi, and A. Viti. 1991. "Central and Peripheral Endocrine Correlates of the Immobility Reaction in the Toad *Bufo bufo*." *Behav. Process* 24:1–7

Luther, D. 2009. "The Influence of the Acoustic Community on Songs of Birds in a Neotropical Rain Forest." *Behav. Ecol* 20:864–871.

Luther D., and L. Baptista. 2010. "Urban Noise and the Cultural Evolution of Bird Songs." *Proc. R. Soc.* B 277:469–473.

Luther D., and K. Gentry. 2013. "Sources of Background Noise and Their Influence on Vertebrate Acoustic Communication." *Behaviour* 150:1045–1068.

Lyn, Heidi, Jamie L. Russell, David A. Leavens, Kim A. Bard, Sarah T. Boysen, Jennifer A. Schaeffer, and William D. Hopkins. 2014. "Apes Communicate about Absent and Displaced Objects: Methodology Matters." *Animal Cognition* 17(1):85–94.

Lynas, Mark. 2011. *The God Species. Saving the Planet in the Age of Humans*. Washington: National Geographic.

Lynas, Mark. 2015. "A Good Anthropocene?—Speech to Breakthrough Dialogue 2015." Accessed July 13, 2015. http://www.marklynas.org/2015/06/a-good-anthropocene-speech-to-breakthrough-dialogue-2015/.

MacWhinney, Brian. 2008. "How Mental Models Encode Embodied Linguistic Perspectives." In *Embodiment, Ego-Space, and Action*, edited by Roberta L. Klatzky, Brian MacWhinney, and Marlene Behrmann, 365–405. New York: Psychology Press.

Maddison, Angus 2003. *World Economy: Historical Statistics*. Paris: OECD.

Magrath, Robert D., Tonya M. Haff, Pamela M. Fallow, and Andrew N. Radford. 2015. "Eavesdropping on Heterospecific Alarm Calls: From Mechanisms to Consequences." *Biological Reviews* 90(2):560– 86.

Malm, Andreas. 2015. "The Anthropocene Myth." *Jacobin*, March 30.

Malm, Andreas, and Alf Hornborg. 2014. "The Geology of Mankind? A Critique of the Anthropocene Narrative." *The Anthropocene Review* 1(1):62–69. doi:10.1177/2053019613516291.

Maran, Timo. 2012. "Are Ecological Codes Archetypal Structures?" In *Semiotics in the Wild: Essays in Honour of Kalevi Kull on the Occasion of his 60th Birthday*, edited by Timo Maran, Kati Lindström, Riin Magnus, and Morten Tønnessen. Tartu: University of Tartu Press.

Maran, Timo. 2015. "Emergence of the 'Howling Foxes': A Semiotic Analysis of Initial Interpretations of the Golden Jackal (*Canis aureus*) in Estonia." *Biosemiotics* 2015. Accessed July 13, 2015. doi:10.1007/s12304-015-9244-1.

Marder, Michael. 2012a. "The Life of Plants and the Limits of Empathy." *Dialogue* 51:259–273.

Marder, Michael. 2012b. "Plant Intentionality and the Phenomenological Framework of Plant Intelligence." *Plant Signaling & Behavior* 7:1365–1372.

Marder, Michael. 2013. "What is Plant-Thinking?" *Klesis—revue philosophique* 25:124–143.

Marder, Michael. 2014. "For a Phytocentrism to Come." *Environmental Philosophy*. Accessed June 20, 2015. doi:10.5840/envirophil20145110.

Margono, Belinda M., Peter V. Potapov, Svetlana Turubanova, Fred Stolle, and Matthew C. Hansen. 2014. "Primary Forest Cover Loss in Indonesia over 2000–2012." *Nature Climate Change* 4:730–735. doi:10.1038/nclimate2277.

Margulis, Lynn. 1998. *The Symbiotic Planet: A New Look at Evolution*. London: Weidenfeld & Nicolson.

Margulis, Lynn, and John L. Hall. 2011. "From Movement to Sensation." In *Chimeras and Consciousness: Evolution of the Sensory Self*, edited by Lynn Margulis, Celeste A. Asikainen, and Wolfgang E. Krumbein, 159–66. Cambridge: MA: MIT Press.

Margulis, Lynn, Celeste A. Asikainen, and Wolfgang E. Krumbein. 2011. "Life's Sensibilities." In *Chimeras and Consciousness: Evolution of the Sensory Self*, edited by Lynn Margulis, Celeste A. Asikainen, and Wolfgang E. Krumbein, 1–13. Cambridge: MA: MIT Press.

Marino, Lori. 2013. "Humans, Dolphins, and Moral Inclusivity." In *The Politics of Species: Reshaping our Relationship with Other Animals*, edited by Raymond Corbey, and Annette Lanjouw, 95–105. Cambridge: Cambridge University Press.

Marks, Jonathan. 2002. *What it Means to be 98% Chimpanzee: Apes, People and their Genes.* Berkeley, CA: University of California Press.

Marsh, James. 2011. *Project Nim*. (Documentary Film). BBC Films, Passion Pictures, and Red Box Films. DVD released 7 February 2012 by Lionsgate Home Entertainment.

Marshack, Alexander. 1972. *The Roots of Civilization*. New York: McGraw-Hill.

Marshack, Alexander. 1975. "Exploring the Mind of Ice-Age Man." *National Geographic* 147:64–89.
Marten K., and Marler P. 1977. "Sound Transmission and Its Significance for Animal Vocalization." *Behav. Ecol. Sociobiol* .2:271–290.
Martin, Diane. 2014. Zen Roshi. Lecture.
Martin, R. R. George. 1979. "Sandkings." *Omni*, 2(8).
Marx, Leo. 1964. *The Machine in the Garden: Technology and the Pastoral Ideal in America.* New York: Oxford University Press.
Maynard Smith, John, and Eörs Szathmáry. 1995. *The Major Transitions in Evolution.* Oxford: Oxford University Press.
Mayr, Ernst. 2004. *What Makes Biology Unique? Considerations on the Autonomy of a Scientific Discipline.* Cambridge: Cambridge University Press.
Mazoyer, Marcel, and Laurence Roudart. 2006. A History of World Agriculture: From the Neolithic Age to the Current Crisis. London, UK: Earthscan.
McAdam, Colin. 2014. *A Beautiful Truth.* New York: Soho Press.
McCauley, R. D., J. Fewtrell, and A. N. Popper. 2003. "High Intensity Anthropogenic Sound Damages Fish Ears." *J. Acoust. Soc. Am.* 113(1):638–642.
McComb, K., D. Reby, L. Baker, C. Moss, and S. Sayalel. 2003. "Long Distance Communication of Acoustic Cues to Social Identity in African Elephants." *Animal Behaviour* 65:317–329.
McDonald, M. A., J. A. Hildebrand, and S. M. Wiggins. 2006. "Increases in Deep Ocean Ambient Noise in the Northeast Pacific West of San Nicolas Island, California." *J. Acoust. Soc. Am.* 120(2):711–718.
McFarland, David. 1999. *Animal Behaviour.* Edinburgh: Pearson Education.
McGhee, George R. 2011. *Convergent Evolution: Limited Forms Most Beautiful.* Cambridge, Mass., London: MIT Press.
McKibben, Bill. 1989. *The End of Nature.* New York: Random House.
McKibben, Bill. 2003. *The End of Nature.* London: Bloomsbury.
McLaughlin, K. E., and H. P. Kunc. 2012. "Experimentally Increased Noise Levels Change Spatial and Singing Behaviour." *Biol. Lett.* 9:20120771.
McNiven, Ian J. 2010. "Navigating the Human-Animal Divide: Marine Mammal Hunters and Rituals of Sensory Allurement." *World Archaeology* 42(2): 215–230.
Meijer, Eva. 2013. "Political Communication with Animals." *Humanimalia: A Journal of Human/Animal Interface Studies* 5:1.
Merchan C. I., L. Diaz-Balteiro, and M. Solino. 2014. "Noise Pollution in National Parks: Soundscape and Economic Valuation." *Landscape and Urban Planning* 123:1–9.
Merchant, Carolyn. 1980. *The Death of Nature: Women, Ecology, and the Scientific Revolution.* San Francisco: Harper and Row.
Merleau-Ponty, Maurice. 1945. *Phénoménologie de la perception.* Paris: Gallimard.
Merleau-Ponty, Maurice. 1961. *L'Œil et l'esprit.* Paris: Gallimard.
Merleau-Ponty, Maurice. 1962. *Phenomenology of Perception.* London: Routledge.
Merleau-Ponty, Maurice. 1963. *The Structure of Behavior.* Boston: Beacon Press.
Merleau-Ponty, Maurice. 1964. *Le visible et l'invisible.* Paris: Gallimard.
Merleau-Ponty, Maurice. 1968. *The Visible and the Invisible.* Evanston, IL: Northwestern University Press.
Merleau-Ponty, Maurice. 1995. *La Nature. Notes de cours au Collège de France* (1957-1960). Paris: Seuil.
Merleau-Ponty, Maurice. 2003. *Nature: Course Notes from the Collège de France.* Evanston, IL: Northwestern University Press.
Michod, Richard E., and Denis Roze. 1997. "Transitions in Individuality." *Proceedings of the Royal Society B* 264(1383):853–857.
Miksis-Olds, J. L., and T. Wagner. 2011. "Behavioral Response of Manatees to Variations in Environmental Sound Levels." *Marine Mammal Science* 27(1):130–148.
Miles, H. Lyn. 1990. "The Cognitive Foundations for Reference in a Signing Orangutan." In *Language and Intelligence in Monkeys and Apes: Comparative Developmental*

*Perspectives*, edited by Sue Taylor Parker, and Kathleen R. Gibson, 511–539. Cambridge, UK: Cambridge University Press.

Millennium Ecosystem Assessment. 2005. *Ecosystems and Human Well-being: General Synthesis*. Washington, DC: Island Press.

Miller, R. James. 2005. "Biodiversity Conservation and the Extinction of Experience." *Trends Ecol. Evol.* 20(8):430–434. doi:10.1016/j.tree.2005.05.013.

Minderman, Jeroen, Elisa Fuentes-Montemayor, James W. Pearce-Higgins, Chris J. Pendlebury, and Kirsty J. Park. 2015. "Estimates and Correlates of Bird and Bat Mortality at Small Wind Turbine Sites." *Biodiversity and Conservation* 24(3):467–82.

Mitchell, Robert W., ed. 2002. *Pretending and Imagination in Animals and Children*. Cambridge: Cambridge University Press.

Mitchell, Robert W. 2013. "The Comparative Study of Imagination." In *The Oxford Handbook of the Development of Imagination*, edited by Marjorie Taylor. Oxford University Press. Online edition. Accessed Oct. 2013. http://www.oxfordhandbooks.com/view/10.1093/oxfordhb/9780195395761.001.0001/oxfordhb-9780195395761-e-030.

Mockford, E. J., and R. C. Marshall. 2009. "Effects of urban noise on song and response behaviour in great tits." *Proc. R. Soc.* B 276:2979–2985.

Monastersky, Richard. 2015. "Anthropocene: The Human Age." *Nature* 519 (March 12, 2015):144–147.

Monbiot, George. 2013. "A Manifesto for Rewilding the World." *The Guardian*, 28 May 2013.

Monbiot, George. 2013. *Feral: Searching for Enchantment on the Frontiers of Rewilding*. London: Allen Lane.

Monod, Jacques. 1972. *Chance and Necessity: An Essay on the Natural Philosophy of Modern Biology*. London: William Collins Sons & Co.

Moon, Kyungwon, and Joonho Jeon. 2015. *The Ways of Folding Space & Flying*, 2015, HD Film Installation, Korean Pavillion. Venice Biennale 2015.

Moore, Kathleen Dean. 2013. "Anthropocene is the Wrong Word." *Earth Island Journal*, spring 2013.

Morton, E. S. 1975. "Ecological Sources of Selection on Avian Sounds." *Am. Nat.* 109:17–34.

Morton, Timothy. 2010. *The Ecological Thought*. Cambridge, MA: Harvard University Press.

Mouse Genome Sequencing Consortium. 2012. "The Mouse Genome. Initial Sequencing and Comparative Analysis of the Mouse Genome." *Nature* 420: 520–562 (509–587). 5 December. doi:10.1038/nature01262.

Moynihan, M. H. 1991. "Structures of Animal Communication." *Man and Beast Revisited*, edited by Michael H. Robinson, and Lionel Triger, 193–202.

Muir, John. 1911 [1988]. *My First Summer in the Sierra.* Boston: Houghton Mifflin.

Mullin, Molly. 1999. "Mirrors and Windows: Sociocultural Studies of Human-Animal Relationships." *Annual Review of Anthropology* 28:201–224.

Mullin, Molly. 2002. "Animals and Anthropology." *Society and Animals* 10(4):387–393.

Nadasdy, Paul. 2007. "The Gift in the Animal: The Ontology of Hunting and Human–Animal Sociality." *American Ethnologist* 34(1):25–43.

Næss, Arne. 1973. "The Shallow and the Deep, Long-Range Ecology Movement. A summary." *Inquiry* 16(1-4):95–100.

Næss, Arne. 1990. *Ecology, Community and Lifestyle*. Cambridge: Cambridge University Press.

Nagasawa, Miho, Shouhein Mitsui, Shiori En, Nobuyo Ohtani, Mitsuaki Ohta, Yasuo Sakuma, and Takefumi Kikusui. 2015. "Oxytocin-Gaze Positive Loop and the Coevolution of Human-Dog Bonds." *Science*, 348(6232):333–336.

Nandini Mitra, Maureen. 2013. "Extremely Loud: We Have Drowned out the Natural Landscape." *Earth Island Journal*, spring 2013.

Narins, P. M., A. S. Feng, W. Y. Lin, H. U. Schnitzler, A. Denzinger, R. A. Suthers, and C. H. Xu. 2004. "Old World Frog and Bird Vocalizations Contain Prominent Ultrasonic Harmonics." *J. Acoust. Soc. Am.* 115:910–913.

Nash, Roderick. 1967. *Wilderness and the American Mind*. New Haven: Yale University Press.

Nealson, Kenneth H. 2011. "Early sensibilities." In *Chimeras and Consciousness: Evolution of the Sensory Self*, edited by Lynn Margulis, Celeste A. Asikainen, and Wolfgang E. Krumbein, 45–52. Cambridge: MA: MIT Press.

Nicholson, Daniel J. 2013. "Organism =/= Machines." *Studies in History and Philosophy of Biological and Biomedical Sciences* 44:669–678.

Nicholson, Daniel J. 2014. "The Return of the Organism as a Fundamental Explanatory Concept in Biology." *Philosophy Compass* 9(5):347–359.

Norment, Camille. 2015. *ŒRapture. Installation*. The Nordic Pavillion, Giardini, Venice Biennale.

Nöth, Winfried. 2009. "On the Instrumentality and Semiotic Agency of Signs, Tools, and Intelligent Machines." *Cybernetics and Human Knowing* 16(3-4):11–36.

Nottebohm, Fernando. 1980. "Brain Pathways for Vocal Learning in Birds." In *Progress in Psychobiology and Physiological Psychology*, edited by James M. Sprague, and Alan N. Epstein, 86–125. New York: Academic Press.

O'Connor, P. 2008. "Sound of Silence: Valuing Acoustics in Heritage Conservation." *Geographical Research* 46(3):361–373.

O'Connor, Terry P. 1997. "Working at Relationships: Another Look at Animal Domestication." *Antiquity* 71:149–156.

O'Neill, John. 2003. "Meta-ethics." In *A Companion to Environmental Philosophy*, edited by Dale Jamieson, 163–176. Malden: Blackwell Publishing.

O'Neill, John, Alan Holland, and Andrew Light. 2008. *Environmental Values*. New York: Routledge.

Oelschlaeger, Max. 2014. "Deep Ecology and the Future of the Wild in the Anthropocene." *The Trumpeter* 30(2):231–246.

Okasha, Samir. 2011. "Biological Ontology and Hierarchical Organization: A Defense of Rank Freedom." In *The Major Transitions in Evolution Revisited*, edited by Brett Calcott and Kim Sterelny, 53–63. Cambridge, MA: MIT Press.

Oliver, Kelly. 2007. "Stopping the Anthropological Machine: Agamben with Heidegger and Merleau-Ponty." *PhaenEx* 2:1–23.

Oliver, Kelly. 2010. "Animal Ethics: Toward an Ethics of Responsiveness." *Research in Phenomenology* 40(2):267–80.

Olwig, Kenneth. 1996. "Reinventing Common Nature: Yosemite and Mt. Rushmore." In *Uncommon Ground: Rethinking the Human Place in Nature*, edited by William Cronon, 379–408. New York: W.W. Norton.

Orr, David. 1999. "Verbicide." *Conservation Biology* 13(4):696.

Ortega, C. P. 2012. "Effects of Noise Pollution on Birds." *Ornithological Monographs* 74(1):6–22.

Palmer, Joy A., Jennifer Suggate, Barbara Bajd, Paul Hart, Roger K. P. Ho, J. K. W. Ofwono-Orecho, Marjorie Peries, Ian Robottom, Elissavet Tsaliki, and Christie Van Staden. 1998. "An Overview of Significant Influences and Formative Experiences on the Development of Adults' Environmental Awareness in Nine Countries." *Environmental Education Research* 4: 445–64. doi:10.1080/1350462980040408.

"Parent Trap." *This American Life* (radio program). 19 February, 2010. http://www.thisamericanlife.org/radio-archives/episode/401/parent-trap.

Parks, S. E., C. W. Clark, and P. L. Tyack. 2007. "Short and Long-Term Changes in Right Whale Calling Behavior: The Potential Effects of Noise on Acoustic Communication." *J. Acoust. Soc. Am.* 122(6):3725–3731.

Parsons, E. C. M. 2012. "From Whaling to Whale Watching: A History of Cetaceans in Scotland." *The Glasgow Naturalist* 25(4). http://www.glasgownaturalhistory.org.uk/gn25_4/parsons.pdf. Accessed October 29 2014.

Paton, L., F. Romero, J. Cuenca, and J. C. Escudero. 2012. "Tolerance to Noise in 91 Bird Species from 27 Urban Gardens of Iberian Peninsula." *Landscape and Urban Planning* 104:1–8.

Patterson, Francine, and Eugene Linden. 1981. *The Education of Koko*. New York: Holt Rinehart & Winston.

Patterson, Francine, and Eugene Linden. 1982. *The Education of Koko*. Andre Deutsch, London.

Peacock, Doug. 1990. *Grizzly Years: In Search of American Wilderness*. New York: Henry Holt and Company, LLC

Penn, Derek C., and Daniel J. Povinelli. 2007. "On the Lack of Evidence that Non-Human Animals Possess Anything Remotely Resembling a 'Theory of Mind'." *Philosophical Transactions of the Royal Society B* 362:731–744. doi:10.1098/rstb.2006.2023.

Penn, Derek C., Keith J. Holyoak, and Daniel J. Povinelli. 2008. "Darwin's Mistake: Explaining the Discontinuity Between Human and Nonhuman Minds." *Behavioral and Brain Sciences* 31:109–178.

Pepper, John W., and Matthew D. Herron. 2008. "Does Biology Need an Organism Concept?" *Biological Reviews* 83:621–627.

Pepperberg, Irene M. 1987. "Interspecies Communication: A Tool for Assessing Conceptual Abilities in the African Grey Parrot (*Psittacus Erithacus*)." In *Language, Cognition, Consciousness: Integrative Levels*, edited by Gary Greenberg and Ethel Tobach, 31–56. Hillsdale, NJ: Erlbaum.

Pepperberg, Irene. 1995. "Grey Parrot Intelligence." *Proceedings of the International Aviculturists Society*. January, 11–15.

Phillips, C. J. C. 2009. "Housing, Handling and the Environment for Cattle." *Principles of Cattle Production*, 95–128.

Pijanowski, B. C., L. J. Villanueva-Rivera, S. L. Dumyahn, A. Farina, B. Krause, B. M. Napoletano, S. H. Gage, and N. Pieretti. 2011. "Soundscape ecology: the science of sound in the landscape." *BioScience* 61(3):203–216.

Pika, Simone, and Thomas Bugnyar. 2011. "The Use of Referential Gestures in Ravens (*Corvus corax*) in the Wild." *Nature Communications* 2:560.

Pilcher, E. J., P. Newman, and R. E. Manning. 2009. "Understanding and Managing Experiential Aspects of Soundscapes at Muir Woods National Monument." *Environmental Management* 43:425–435.

Pilley, John, and Allison Reid. 2011. "Border Collie Comprehends Object Names as Verbal Referents." *Behavioural Processes*, 86(2):184–195.

Pilley, John. 2013. "Border Collie Comprehends Sentences Containing a Prepositional Object, Verb, and Direct Object." *Learning and Motivation* 44(4):229–240.

Pine, M. K., A. G. Jeffs, and C. A. Radford. 2013. "The Cumulative Effect on Sound Levels from Multiple Underwater Anthropogenic Sound Sources in Shallow Coastal Waters." *Journal of Applied Ecology* 51:23–30.

Pinker, Steven. 1997. *How the Mind Works*. New York: Norton.

Pivato, Stefano. 2011. *Il secolo del rumore*. Il Mulino, Milano.

Plessner, Helmuth. 1975. *Die Stufen des Organischen und der Mensch. Einleitung in die philosophische Anthropologie*. Berlin: de Gruyter.

Plessner, Helmuth. 1982. *Mit anderen Augen. Aspekte einer philosophischen Anthropologie*. Stuttgart: Reclam.

Plumwood, Val. 1996, "Nature, Self, and Gender: Feminism, Environmental Philosophy, and the Critique of Rationalism." In *Ecological Feminist Philosophies*, edited by Karen J. Warren, 155–180. Bloomington: Indiana University Press.

Plumwood, Val. 2000. "Being Prey." In *The Ultimate Journey: Inspiring Stories of Living and Dying*, edited by James O'Reilly, Sean O'Reilly, and Richard Sterling, 128–146. San Francisco: Travelers' Tales.

Plumwood, Val. 2012. "The Wisdom of the Balanced Rock: The Parallel Universe and the Prey Perspective." In *The Eye of the Crocodile*, edited by Shannon Lorraine, 35–45. Canberra: Australian National University E Press. http://press.anu.edu.au/?p=208511.

Pollan, Michael. 2013. "The Intelligent Plant." *The New Yorker*, December 23. Accessed July 6, 2015. http://www.newyorker.com/magazine/2013/12/23/the-intelligent-plant.

Portmann, Adolf, and Richard B. Carter. 1990. *Essays in Philosophical Zoology by Adolf Portmann. The Living Form and the Seeing Eye*. Lewiston: The Edwin Mellen Press.

Portmann, Adolf. 1958. "Selbstdarstellung als Motiv der lebendigen Formbildung." *Geist und Werk. Aus der Werkstatt unserer Autoren.* Zurn 75, Geburtstag von Dr. Daniel Brody. Rhein Verlag: Zurich.

Portmann, Adolf. 1965. *Die Tiergestalt*. Freiburg, Basel, Wien: Herder. English translation by H. Czech, New York: Schocken Books, 1967.

Portmann, Adolf. 1973. *Biologie und Geist*. Frankfurt am Main: Suhrkamp.

Potash, LM. 1972. "Noise-Induced Changes in Calls of the Japanese Quail." *Psychon. Sci.* 26:252–254.

Prather, Jonathan F., S. Peters, Stephen Nowicki, and R. Mooney. 2008. "Precise Auditory-Vocal Mirroring in Neurons for Learned Vocal Communication." *Nature* 451:305–310.

Prigogiene, Ilya, and Isabelle Stengers. 1984. *Order out of Chaos: Man's New Dialogue with Nature.* New York: Bantam Books.

Proppe, D, C. B. Sturdy, and C. Cassady St. Clair. 2013. "Anthropogenic Noise Decreases Urban Songbird Diversity and May Contribute to Homogenization." *Global Change Biology* 19:1075–1084.

Pryor, Francis. 2003. *Britain B.C.: Life in Britain and Ireland Before the Romans.* New York: Harper.

Rabin, L. A., B. McCowan, S. L. Hooper, and D. H. Owings. 2003. "Anthropogenic Noise and Its Effect on Animal Communication: An Interface between Comparative Psychology and Conservation Biology." *Int. J. Comparative Psychology* 16:172–192.

Rabin, L. A., R. G. Goss, and D. H. Owings. 2006. "The Effects of Wind Turbines on Antipredator Behavior in California Ground Squirrels (*Spermophlus Beecheyi*)." *Biological Conservation* 131:410–420.

Radford, A. N., E. Kerridge, and S. D. Simpson. 2014. "Acoustic Communication in a Noisy World: Can Fish Compete with Anthropogenic Noise?" *Behavioural Ecology.* doi:10.1093/beheco/aru029

Radford, C. A., J. A. Stanley, C. T. Tindle, J. C. Montgomery, and A. G. Jeffs. 2010. "Localised Coastal Habitats Have Distinct Underwater Sound Signatures." *Marine Ecology Progress Series* 401:21–29.

Rankin, Aidan. 2009. "Karmic Ecology: Lessons from the Jain Dharma." *The Trumpeter Journal of Ecosophy* 25(2):1–14.

Read, J., G. Jones, and A. N. Radford. 2013. "Fitness Costs As Well As Benefits Are Important When Considering Responses to Anthropogenic Noise." *Behavioural Ecology* 25(1):4–7.

Rebanks, James. 2015. *The Shepherd's Life: Modern Dispatches from an Ancient Landscape.* New York: Flatiron Books.

Reiner, Anton, David J. Perkel, Laura L. Bruce, Ann B. Butler, András Csillag, Wayne Kuenzel, Loreta Medina, et al. 2004. "Revised Nomenclature for Avian Telencephalon and Some Related Brainstem Nuclei." *Journal of Comparative Neurology* 473(3):377–414.

Renner, Michael, Sean Sweeney, and Jill Kubit. 2008. *Green Jobs. Working for People and the Environment.* Worldwatch Report 177. Washington DC: Worldwatch Institute.

Ricoeur, Paul. 1973. "The Model of the Text: Meaningful Action Considered as a Text." *New Literary History* 5(1):91–117. doi:10.2307/468410.

Ricoeur, Paul. 1981. *Hermeneutics and the Human Sciences.* Translated by John B. Thompson. Cambridge: Cambridge University Press.

Ricoeur, Paul. 1991. *From Text to Action. Essays in Hermeneutics II.* Translated by Kathleen Blamey and John B. Thompson. Evanston: Northwestern University Press.

Ricoeur, Paul. 1992. *Oneself as Another.* Translated by Kathleen Blamey. Chicago: University of Chicago Press.

Riis, Søren. 2011. "Towards the Origin of Modern Technology: Reconfiguring Martin Heidegger's Thinking." *Continental Philosophical Review* 44:103–117. doi:10.1007/s11007-011-9170-0.

Riley, Christopher. 2014. The Girl Who Talked to Dolphins. TV movie. Production company: Shadow Industries. Shown on BBC Four on June 17, 2014. IMDB: http://www.imdb.com/title/#3822614/.

Rindos, David. 1984. *The Origins of Agriculture: An Evolutionary Perspective.* New York: Academic Press.

Robbins, C. S. 1981. "Bird Activity Levels Related to Weather." In *Estimating Numbers of Terrestrial Birds,* edited by J. Ralph, and M. J. Scott, 301–310. Allen Press, Inc. Studies in Avian Biology No. 6. Lawrence, Kansas.

Rogers, Lesley J., and Gisela Kaplan. 2004. "All Animals Are *Not* Equal: The Interface between Scientific Knowledge and the Legislation for Animal Rights." In *Animal Rights: Law and Policy,* edited by Cass R. Sunstein, and Martha C. Nussbaum, 175–204. Oxford: Oxford University Press.

Rogers, Lesley J., and Gisela Kaplan 2006. "Think or be Damned: Problematic Case of Higher Cognition in Animals and Legislation for Animal Welfare." *Journal of Animal Law* 12(2):151–191.

Rose, Deborah Bird. 2011. *Wild Dog Dreaming: Love and Extinction.* Charlottesville: University of Virginia Press.

Rose, Hilary, and Steven Rose, eds. 2000. *Alas, Poor Darwin: Arguments Against Evolutionary Psychology.* New York: Harmony Books.

Rose, Michael R., and Todd H. Oakley. 2007. "The New Biology: Beyond the Modern Synthesis." *Biology Direct* 2(30).

Rose, Steven. 1997. *Lifelines: Biology, Freedom, Determinism.* London: The Penguin Press.

Rothenberg, David. "Whale Song Explained." https://medium.com/@dealville/whales-synchronize-their-songs-across-oceans-and-theres-sheet-music-to-prove-it-b1667f603844, 2014. Accessed September 29, 2015.

Rothwell, Jerry. 2015. How to Change the World. Co-production by Met Film, Insight Productions and Daniel Film. More info on IMDB: http://www.imdb.com/title/#4144504/.

Rowe, Candy, and Susan D. Healy. 2014. "Measuring Variation in Cognition." *Behavioral Ecology* 25(6):1287–1292. doi:10.1093/beheco/aru090.

Rowley-Conwy, Peter, and Robert Layton. 2011. "Foraging and Farming as Niche Construction: Stable and Unstable Adaptations." *Phil. Trans. R. Soc. B* 366(1566):849–862. doi:10.1098/rstb.2010.0307.

Ruddiman, W. F. 2003. "The Anthropogenic Greenhouse Era began Thousands of Years Ago." *Climatic Change* 61: 261–293.

Russell, Nerissa. 2012. *Social Zooarchaeology: Humans and Animals in Prehistory.* Cambridge: Cambridge University Press.

Russon, Anne, Kim A. Bard, and Sue Parkers, eds. 1996. *Reaching Into Thought: The Minds of the Great Apes.* Cambridge: Cambridge University Press.

Ruyer, Raymond. 1953. "Les conceptions nouvelles de l'instinct." *Les Temps Modernes* N°96, November 1953.

Sachs, Carl B., and Shane Epting. 2010. "Thinking of the Living Body in Hans Jonas and Merleau-Ponty." Accessed June 26, 2015. https://scholar.google.com/scholar?oi=bibs&cluster=8191190888348748898&btnI=1.

Sandberg, Anders. (n.d.). "We borg: Speculations on Hive Minds As a Posthuman State." http://www.aleph.se/Trans/Global/Posthumanity/WeBorg.html. Accessed August 28, 2015.

Savage-Rumbaugh, E. Susan. 1986. *Ape Language: From Conditioned Response to Symbol.* New York: Columbia University Press.

Savage-Rumbaugh, Sue. 1998. "Bringing Up Kanzi." In *Apes, Language, and the Human Mind,* by E. Sue Savage-Rumbaugh, Stuart G Shanker, and Talbot J. Taylor, 3–74. Oxford: Oxford University Press.

Savage-Rumbaugh, Sue. 2004. "The Gentle Genius of Bonobos." http://www.ted.com/talks/susan_savage_rumbaugh_on_apes_that_write.

Savage-Rumbaugh, E. Susan, and Roger Lewin. 1994. *Kanzi: The Ape at the Brink of the Human Mind.* New York: John Wiley & Sons.

Scamardella, Joseph M. 1999. "Not Plants or Animals: A Brief History of the Origin of Kingdoms Protozoa, Protista and Protoctista." *International Microbiology* 2:207–16.

Scheler, Max. 2009. *The Human Place in the Cosmos.* Evanston: Northwestern University Press.

Schmidt, Charlie. 2015. "Alarm over Sinking Delta." *Science* (Weekly News) 348(6237):845–846, 22 May. doi:10.1126/science.348.6237.845.

Schmidt, R., A. Morrison, P. H. Kunc. 2014. "Sexy Voices—No Choices: Male Song in Noise Fails to Attract Females." *Animal Behavior* 94: 55–59.

Schofield, Malcolm. 1995. "Aristotle on the Imagination." In *Essays on Aristotle's De Anima*, edited by Martha C. Nussbaum and Amélie Oksenberg Rorty, 248–277. Oxford: Clarendon Press.

Schwartz, J. J., and K. D. Wells. 1983. "An Experimental Study of Acoustic Interference between Two Species of Neotropical Treefrog with a Complex Vocal Repertoire." *Anim. Behav.* 42:565–577.

Scott, N. J., and E. C. M. Parsons. 2005. "A Survey of Public Opinion in South-West Scotland on Cetacean Conservation Issues." *Aquatic Conservation* 15:299–312.

Seemann, Axel, ed. 2011. *Joint Attention. New Developments in Psychology, Philosophy of Mind and Social Neuroscience.* Cambridge, Mass.: MIT press.

SEO/Birdlife. 2012. "Spanish Wind Farms Kill 6 to 18 Million Birds & Bats a Year." Media Release. http://www.epaw.org/press/STEI_media_release_14Jan2012.pdf.

Seppänen, J., and M. Nieminen. 2004. "Measurements and Descriptions of Underwater Noise in Finland." *Geophysica* 40(1-2):23–38.

Serres, Michel. 1995. *The Natural Contract,* translated by Elizabeth MacArthur, and William Paulson. Ann Arbor, MI: University of Michigan Press.

Serres, Michel. 2008. *The Five Senses: A Philosophy of Mingled Bodies (I),* translated by Margaret Sankey, and Peter Cowley. London: Continuum.

Servais, Véronique. 2005. "Enchanting and Enchanted Dolphins. An Analysis of Human/Dolphin Encounters." In *Animals in Person: Cultural Perspectives on Human-Animal Intimacies,* edited by John Knight. Oxford: Berg.

Shannon, G., L. M. Angeloni, G. Wittemyer, and K. M. Fristrupp. 2014. "Road Traffic Noise Modifies Behaviour of a Keystone Species." *Animal Behavior* 94:135–141.

Shaw, Rachel C., Neeltje J. Boogert, Nicola S. Clayton, and Kevin C. Burns. 2015. "Wild Psychometrics: Evidence for 'General' Cognitive Performance in Wild New Zealand Robins, *Petroica Longipes.*" *Animal Behaviour* 109:101–111. doi.org/10.1016/j.anbehav.2015.08.001.

Sheets-Johnstone, Maxine. 1999. *The Primacy of Movement.* Philadelphia: John Benjamins.

Shettleworth, Sara J. 2009. "Cognition: Theories of Mind in Animals and Humans." *Nature* 459(7246) 506.

Shiva, Vandana. 2013. "Everything I Need to Know I Learned in the Forest." *Yes Magazine* 64(winter 2012):46–50.

Sickler, Jessica, John Fraser, Thomas Webler, Diana Reiss, Paul Boyle, Heidi Lyn, Katherine Lemcke, and Sarah Gruber. 2006. "Social Narratives Surrounding Dolphins: Q Method Study." *Society and Animals* 14(4):351–382.

Simondon, Gilbert. 1958. *Du mode d'existence des objets techniques.* Paris: Aubier.

Simpson, Carl. 2011 "How Many Levels Are There? How Insights from Evolutionary Transitions in Individuality Help Measure the Hierarchical Complexity of Life." In *The Major Transitions in Evolution Revisited,* edited by Calcott, Brett and Kim Sterelny, 200–25. Cambridge, MA: MIT Press.

Simpson, S. D., A. Jeffs, J. C. Montgomery, R. D. McCauley, and M. G. Meekan. 2008. "Nocturnal Relocation of Adult and Juvenile Coral Reef Fishes in Response to Reef Noise." *Coral Reefs* 27:97–104.

Simpson, S. D., P. L. Mumday, M. L. Wittenrich, R. Manassa, D. L. Dixson, M. Gagliani, and H. Y. Yan. 2011. "Ocean Acidification Erodes Crucial Auditory Behaviour in a Marine Fish." *Biol Lett* 7:917–920.

Simpson, S. D., A. N. Radford, E. J. Tickle, M. G. Meekan, and A. G. Jeffs. 2012. "Adaptive Avoidance of Reef Noise." *PLoS ONE* 6(2): e16625. doi:10.1371/journal.pone.0016625.

Singer, Jonathan D. 2015. "'The Flesh of My Flesh': Animality, Difference, and 'Radical Community' in Merleau-Ponty's Philosophy." In *Animal Ethics and Philosophy: Questioning the Orthodoxy*, edited by Elisa Aaltola, and John Hadley, 99–116. New York: Rowman and Littlefield.

Skogen, Ketil, Isabelle Mauz, and Olve Krange. 2008. "Cry Wolf! Narratives of Wolf Recovery in France and Norway." *Rural Sociology* 73(1):105–33.

Slabbekoorn, H. 2004. "Habitat-Dependent Ambient Noise: Consistent Spectral Profiles in Two African Forest Types." *J. Acoust. Soc. Am.* 116(6):3727–3733.

Slabekoorn, H., M. Peet. 2003. "Birds Sing at a Higher Pitch in Urban Noise." *Nature* 424:267.

Slabekoorn, H., and E. A. P. Ripmeester. 2008. "Birdsong and Anthropogenic Noise: Implication and Application for Conservation." *Molecular Ecology* 17:72–83.

Slabbekoorn, H., N. Bouton, I. van Opzeeland, et al. 2010. "A Noisy Spring: The Impact of Globally Rising Underwater Sound Levels on Fish." *Trends. Ecol. Evol.* 25:419–27.

Slobodchikoff, Constantine Nicholas, Bianca S. Perla, and Jennifer L. Verdolin. 2009. *Prairie Dogs: Communication and Community in an Animal Society*. Cambridge: Harvard University Press.

Smallwood, K. Shawn. 2007. "Estimating Wind Turbine-Caused Bird Mortality." *Journal of Wildlife Management* 71(8):2781–91. doi:10.2193/2007-006

Smil, Vaclav. 2011. "Harvesting the Biosphere: The Human Impact." *Population and Development Review* 37(4):613–36.

Smuts, Barbara. 1987. *Primate Societies*. Chicago: University of Chicago Press.

Smuts, Barbara. 1999. "Barbara Smuts" in Appendix, "Reflections." *The Lives of Animals*, by J. M. Coetzee, 107–120. Princeton, NJ: Princeton University Press.

Smuts, Barbara. 2001. "Encounters with Animal Minds." *Journal of Consciousness Studies* 8(5):293–309.

Smuts, Barbara. 2006. "Between Species: Science and Subjectivity." *Configurations*, 14:115–126.

Snyder, Gary. 1990. *The Practice of the Wild*. San Francisco: North Point Press.

Soha, Jill A., and Susan Peters. 2015. "Vocal Learning in Songbirds and Humans: A Retrospective in Honor of Peter Marler." *Ethology* 121(10):933–945. doi:10.1111/eth.12415.

Soltis, Joseph, Lucy E. King, Iain Douglas-Hamilton, Fritz Vollrath, and Anne Savage. 2014. "African Elephant Alarm Calls Distinguish between Threats from Humans and Bees." *PLoS ONE* 9(2): e89403.

Soulé, Michael, and Reed Noss. 1998. "Rewilding and Biodiversity as Complementary Goals for Continental Conservation." *Wild Earth* 22:2–11.

Spelke, Elizabeth S., and Lee Sang Ah. 2012. "Core Systems of Geometry in Animal Minds." *Philosophical Transactions of the Royal Society London B Biological Science* 367(1603):2784–93. doi:10.1098/rstb.2012.0210.

Spencer, Herbert. 1899. *The Principles of Biology: Vol. 1 & 2*. London: Williams & Norgate.

Spinoza, Baruch. 2000. *Ethics*. Edited and translated by G.H.R Parkinson. Oxford: Oxford University Press.

Stapledon, Olaf. 1937. *Starmaker*. London: Methuen.

Steffen, W., A. Sanderson, P. D. Tyson, J. Jäger, P. A. Matson, B. Moore III, F. Oldfield, K. Richardson, H. J. Schellnhuber, B. L. Turner, and R. J. Wasson. 2004. *Global Change and the Earth System: A Planet Under Pressure*. Berlin, Heidelberg, New York: Springer-Verlag.

Steffen, Will, Jacques Grinevald, Paul Crutzen, and John McNeill. 2011. "The Anthropocene: Conceptual and Historical Perspectives." *Phil. Trans. R. Soc. A* 369(1938):842–867. doi:10.1098/rsta.2010.0327.

Steinfeld, Henning, Pierre Gerber, Tom Wassenaar, Vincent Castel, Mauricio Rosales, and Cees de Haan. 2006. *Livestock's Long Shadow. Environmental Issues and Options.* Food and Agriculture Organization of the United Nations FAO.

Sterner, Dave. 2002. "A Roadmap for PIER Research on Avian Collisions with Wind Turbines in California." PIER Study of the California Energy Commission. California Energy Commission.

Stevens, Thomas King, Amanda M. Hale, Kris B. Karsten, and Tory Bennett. 2013. "An Analysis of Displacement from Wind Turbines in a Wintering Grassland Bird Community." *Biodiversity and Conservation* 22(8):1755–67.

Stiegler, Bernard. 2015. *Symbolic Misery, Volume 2: The Catastrophe of the Sensible.* Oxford: Polity.

Strand, Ginger. 2013. "Beware the Rainmakers." *Earth Island Journal,* spring 2013.

Sueur, J., and Almo Farina. 2015. "Ecoacoutics: The Ecological Investigation and Interpretation of Environment Sound." *Biosemiotics.* doi:10.1007/s12304-015-9248-x.

Suzuki, Toshitaka N. 2014. "Communication about Predator Type by a Bird Using Discrete, Graded and Combinatorial Variation in Alarm Calls." *Animal Behaviour* 87:59–65.

Szathmáry, Eörs, and John Maynard Smith. 1995. "The Major Evolutionary Transitions." *Nature* 374(6519): 227–32.

Szerszynski, Bronislaw. 2012. "The End of the End of Nature: The Anthropocene and the Fate of the Human." *Oxford Literary Review* 34(2):165–84.

Szipl, Georgine, Markus Boeckle, Claudia A.F. Wascher, Michela Spreafico, and Thomas Bugnyar. 2015. "With Whom to Dine? Ravens' Responses to Food-Associated Calls Depend on Individual Characteristics of the Caller." *Animal Behaviour 99:33–42.*

Taylor, Hollis. 2010. "Blowin' in Birdland: Improvisation and the Australian Pied Butcherbird." *Leonardo Music Journal* 20:79–83. doi:10.1162/LMJ_a_00016.

Taylor, Marjorie, ed. 2013a. *The Oxford Handbook of the Development of Imagination.* Oxford University Press. Online edition. Accessed October 2015. http://www.oxfordhandbooks.com/view/10.1093/oxfordhb/9780195395761.001.0001/oxfordhb-9780195395761.

Taylor, Marjorie. 2013b. "Looking Ahead: Some Thoughts About Future Directions." In *The Oxford Handbook of the Development of Imagination,* edited by Marjorie Taylor. Oxford University Press. Online edition. Accessed October 2015. http://www.oxfordhandbooks.com/view/10.1093/oxfordhb/9780195395761.001.0001/oxfordhb-9780195395761-e-037.

Taylor, Paul. 1986. *Respect for Nature: A Theory of Environmental Ethics.* Princeton: Princeton University Press

Tennessen, J. B., S. E. Parks, and T. Langkilde. 2014. "Traffic Noise Causes Physiological Stress and Impairs Breeding Migration Behaviour in Frogs." *Conservation physiology* 2:1–8.

Terborgh, John, and James A. Estes, eds. 2010. *Trophic Cascades: Predators, Prey, and the Changing Dynamic of Nature.* Washington: Island Press.

*The Cambridge Declaration on Consciousness.* Edited by Jaak Panksepp, Diana Reiss, David Edelman, Bruno Van Swinderen, Philip Low, and Christof Koch. http://fcmconference.org/img/CambridgeDeclarationOnConsciousness.pdf. Accessed August 12, 2012.

Thompson, Charis. 2002. "When Elephants Stand for Competing Models of Nature." In *Complexities: Social Studies of Knowledge Practices,* edited by John Law and Annemarie Mol. Durham, NC: Duke University Press.

Thoreau, Henry David. 1862. "Walking." *Atlantic Monthly.* http://www.theatlantic.com/magazine/archive/1862/06/walking/304674/.

Thornton, Alex, Jess Isden, and Joah R. Madden. 2014. "Toward Wild Psychometrics: Linking Individual Cognitive Differences to Fitness." *Behavioral Ecology* (2014): doi:10.1093/beheco/aru095.

Thorp, Thomas. 2014. "Eating Wolves." In *Old World and New World Perspectives in Environmental Philosophy*, edited by Martin Drenthen, and Jozef Keulartz, 175–197. Cham, Heidelberg, New York, Dordrecht, London: Springer.

Tolimieri, N., A. Jeffs, and J. C. Montgomery. 2000. "Ambient Sound As a Cue for Navigation by the Pelagic Larvae of Reef Fishes." *Marine Ecology Progress Series* 207:219–224.

Tønnessen, Morten 2008. "The Statistician's Guide to Utopia: The Future of Growth." *TRAMES* 12(62/57), 2:115–126.

Tønnessen, Morten 2010. "Global Species." *New Formations: A Journal of Culture/Theory/Politics* 69:98–110.

Tønnessen, Morten. 2010. "Is a Wolf Wild As Long As It Does Not Know That It Is Being Thoroughly Managed?" *Humanimalia*, 2(1). http://www.depauw.edu/humanimalia/issue03/tonnesen.html. Accessed June 4, 2015.

Tønnessen, Morten. 2010. "Wolf Land." *Biosemiotics* 3(3):289–297. Accessed October 15, 2015. doi:10.1007/s12304-010-9077-x.

Tønnessen, Morten. 2011. *Umwelt transition and Uexküllian phenomenology: An ecosemiotic analysis of Norwegian wolf management*. Doctoral dissertation. University of Tartu, Department of Semiotics.

Tønnessen, Morten. 2015. "Introduction: The Relevance of Uexküll's Umwelt Theory Today." In *Jakob von Uexküll: The Discovery of the Umwelt Between Biosemiotics and Theoretical Biology* (Biosemiotics 9), edited by Carlo Brentari, 1–20. Dordrecht: Springer.

Tønnessen, Morten, and Jonathan Beever. 2015. "Beyond Sentience: Biosemiotics as Foundation for Animal Ethics." In *Animal Ethics and Philosophy: Questioning the Orthodoxy*, edited by Elisa Aaltola and John Hadley, 47–62. New York: Rowman and Littlefield.

Trampe, Wilhelm. 2001. "Language and Ecological Crisis: Extracts from a Dictionary of Industrial Agriculture." In *The Ecolinguistics Reader*, edited by Alwin Fill and Peter Mulhausler. London: Continuum, 232–246.

Truax, Barry, ed. 1999. *Handbook for Acoustic Ecology*. Cambridge, UK: Cambridge Street Publishing.

Tsing, Anna. "Earth Stalked by Man." Paper presented at the 5th Annual Marilyn Strathern Lecture, University of Cambridge, May 21, 2015.

Tsing, Anna. 1995. "Empowering Nature, or: Some Gleanings in Bee Culture." In *Naturalizing Power: Essays in Feminist Cultural Analysis*, edited by Sylvia Yanagisako and Carol Delaney, 113–143. New York: Routledge.

Tsing, Anna. 2011. "Arts of Inclusion, or How to Love a Mushroom." *Australian Humanities Review*, 50:5–21.

Tsing, Anna. 2012. "Unruly Edges: Mushrooms as Companion Species." *Environmental Humanities* 1:141–154.

Turner, Jack. 1997. *The Abstract Wild*. Tucson: The University of Arizona Press.

Uexküll, Jakob von. 1928. *Theoretische Biologie* (2nd ed.). Berlin: Verlag von Julius Springer.

Uexküll, Jakob von. 1934. *Streifzüge durch die Umwelten von Tieren und Menschen*. Berlin: J. Springer.

Uexküll, Jakob von. 1957. *A Stroll Through the Worlds of Animals and Men*. In *Instinctive Behavior: The Development of a Modern Concept*, edited and translated by Claire H. Schiller, 5–80.

Uexküll, Jakob von. 1982. *The Theory of Meaning*. Special issue of *Semiotica*, with an Introduction by Thure von Uexküll. 42:25–81.

Uexküll, Jakob von. 2010. *A Foray into the Worlds of Animals and Humans, With: A Theory of Meaning*. Minneapolis/London: University of Minnesota Press.

Vallortigara, Giorgio, Cinzia Chiandetti, Rosa Rugani, Valeria A. Sovrano, and Lucia Regolin. 2010. "Animal Cognition." *Wiley Interdisciplinary Reviews: Cognitive Science* 1:882–893. doi:10.1002/wcs.75.

Van den Noortgaete, Francis, and Johan De Tavernier. 2014. "Affected by Nature: A Hermeneutical Transformation of Environmental Ethics." *Zygon*, 49:572–592. doi:10.1111/zygo.12103.

Van Tongeren, Paul. 1994. "Moral Philosophy as a Hermeneutics of Moral Experience." *International Philosophical Quarterly* 34(2):199–214. doi:10.5840/ipq19943423.

Van Tongeren, Paul, and Paulien Snellen. 2014. "How Hermeneutics Might Save the Life of (Environmental) Ethics." In *Interpreting Nature: The Emerging Field of Environmental Hermeneutics*, edited by Forrest Clingerman, Martin Drenthen, Brian Treanor, and David Utsler, 297–312. New York: Fordham University Press.

Van Kenpen, E., I. van Kamp, M. Nilsson, J. Lammers, H. Emmen, C. Clark, and S. Stansfeld. 2010. "The Role of Annoyance in the Relation between Transportation Noise and Children's Health and Cognition." *J. Acoust. Soc. Am.* 128(5):2817–2828.

Vance, Erik. 2009. "High Hopes." *Nature* 460(7255):564–66.

Vermeij, M. J. A., K. L. Marhaver, C. M. Huijbers, I. Nagelkerken, and S. D. Simpson. 2010. "Coral Larvae Move toward Reef Sounds." *PLoS ONE* 5(5):e10660.

Vest, Jay Hansford C. 1985. "Will-of-the-Land: Wilderness among Primal Indo-Europeans." *Environmental Review*, 9(4):323–329. doi:10.2307/3984462.

Wahlberg, M., and H. Westerberg. 2005. "Hearing in Fish and Their Reactions to Sounds from Offshore Wind Farms." *Marine Ecology Progress Series* 288:295–309.

Wale, A. M., D. S. Simpson, and N. A. Radford. 2013. "Noise Negatively Affects Foraging and Antipredaotr Behavior in Shore Crabs." *Animal Behaviour* 86: 111–18.

Walther, G. R., E. Post, P. Convey, A. Menzel, C. Parmesan, T. Beebee, J. M. Fromentin, O. Hoegh-Guldberg, and F. Bairlein. 2002. "Ecological Responses to Recent Climate Change." *Nature* 416:389–395.

Watts, R. D., R. W. Compton, J. H. McCammon, C. L. Rich, S. M. Wright, T. Owens, and D. S. Ouren. 2007. "Roadless Space of the Conterminous United States." *Science* 316:736.

Weingarten Matthew, Shemin Ge, Jonathan W. Godt, Barbara A. Bekins, and J. L. Rubinstein. 2015. "High-rate injection is associated with the increase in U.S. mid-continent seismicity." *Science* 348(6241):1336– 40. doi:10.1126/science.aab1345.

Weisman, Alan. 2007. *The World Without Us*. New York: St. Martin's Press.

Welch, Rick, and James Clegg. 2010. "From Protoplasmic Theory to Cellular Systems Biology: A 150-Year Reflection." *American Journal of Physiology—Cell Physiology* 298(6):1280–1290.

Wengrow, David. 2006. *The Archaeology of Early Egypt. Social Transformations in North East Africa, 10,000 to 2650 BC*. Cambridge: Cambridge University Press.

Wenner, Melinda. 2007. "Humans Carry More Bacterial Cells than Human Ones." *Scientific American*. http://www.scientificamerican.com/article/strange-but-true-humans-carry-more-bacterial-cells-than-human-ones/.

Westling, Louise. 1996. *The Green Breast of the New World: Landscape, Gender, and American Fiction*. Athens, GA: University of Georgia Press.

Westling, Louise. 2014. *The Logos of the Living World: Merleau-Ponty, Animals, and Language*. New York: Fordham University Press.

Westling, Louise. 2014. "The Zoosemiotics of Sheep Herding with Dogs." *The Semiotics of Animal Representations*, edited by Kadri Tüür and Morten Tønnessen, 33–52. Amsterdam: Rodopi.

Westling, Louise. 2016. "Merleau-Ponty and the Eco-Literary Imaginary." In *Handbook of Ecocriticism and Cultural Ecology*, edited by Hubert Zapf. Berlin: DeGruyter.

White, Lynn. 1974. "The Historical Roots of our Ecological Crisis." In *Ecology and Religion in History*, edited by David Spring and Eileen Spring. New York: Harper & Row.

Whitefield, A. K., and Becker A. 2014. "Impacts of Recreational Motorboats on Fishes: A Review." *Marine Pollution Bulletin* 83:24–31.

Whitehead, Hal, and Luke Rendell. 2015. *The Cultural Lives of Whales and Dolphins.* London: University of Chicago Press.

Whittle, Alistair. 2003. *The Archaeology of People. Dimensions of Neolithic life.* London and New York: Routledge.

Wiley, R. H. 1994. "Errors, Exaggeration, and Deception in Animal Communication." In *Behavioral Mechanisms in Evolutionary Ecology,* edited by Leslie A. Real, 157–189. Chicago, IL: University of Chicago Press.

Willers, William B. 1999. Introduction to *Unmanaged Landscapes, Voices for Untamed Nature,* edited by William B. Willers. Washington: Island Press.

Willerslev, Rane. 2007. *Soul Hunters. Hunting, Animism, and Personhood among the Siberian Yukaghirs.* Berkeley, Los Angeles, London: University of California Press.

Willerslev, Rane. 2012. *On the Run in Siberia.* Minneapolis and London: University of Minnesota Press.

Willett, Cynthia. 2014. *Interspecies Ethics.* New York: Columbia University Press.

Williams, Bernard. 1985. *Ethics and the Limits of Philosophy.* London: Fontana.

Williams, R., C. W. Clark, D. Ponirakis, and E. Ashe. 2013. "Acoustic Quality of Critical Habitats for Three Threatened Whale Populations." *Animal Conservation* 17:174–185.

Wilson, Edward O., (March 2016). *Half-Earth: Our Planet´s Fight for Life.* W.W. Norton & Company.

Winder, Virginia, Andrew Gregory, Lance McNew, and Brett Sandercock. 2015. "Responses of Male Greater Prairie-Chickens to Wind Energy Development" *The Condor* 117(2):284.

Withnall, Adam. 2015. "Earth Overshoot Day: Humanity Consumes 100% of Planet's Resources in Eight Months." *The Independent.* http://www.independent.co.uk/news/science/earth-overshoot-day-humanity-consumes-100-of-planets-resources-in-eight-months-10457683.html. Accessed October 29, 2015.

Wittgenstein, Ludwig. 1958. *Philosophical Investigations.* Oxford: Blackwell.

Wittgenstein, Ludwig. 1978. *Lectures and Conversations on Aesthetics, Psychology and Religious Belief.* Oxford: Blackwell.

Woese, Carl R. 2004. "A New Biology for a New Century." *Microbiology and Molecular Biology Reviews* 68(2):173–186.

Wolfe, Cary. 2003. *Animal Rites: American Culture, the Discourse of Species, and Posthumanist Theory.* Chicago: Chicago University Press.

Wolfe, Charles T. 2010. "Do Organisms Have an Ontological Status?" *History and Philosophy of the Life Sciences* 32(2/3):195–231.

Wong, Kate. 2012. "Troubled Ape Facility Reinstates Controversial Researcher." *Scientific American.* November 21. http://blogs.scientificamerican.com/observations/troubled-ape-facility-reinstates-controversial-researcher/.

Woodwell, George M. 2010. "The Biodiversity Blunder." *BioScience* 60:870–871.

Youth, Howard. 2003. *Winged Messengers: The Decline of Birds.* Worldwatch Paper 165. Washington DC: Worldwatch Institute.

Zahavi Amotz. 1977. "The Cost of Honesty (further remarks on the handicap principle)." *J. Theor. Biol* 67:603–605.

Zalasiewicz, Jan, Mark Williams, Alan Smith, Tiffany L. Barry, Paul R. Bown, Peter Rawson, Patrick Brenchley, David Cantrill, Angela L. Coe, Andrew Gale, Philip L. Gibbard, F. John Gregory, Mark W. Hounslow, Andrew C. Kerr, Paul Pearson, Robert Knox, John Powell, Colin Waters, John Marshall, Michael Oates, and Philip Stone. 2008. "Are We Now Living in the Anthropocene?" *GSA Today* 18(2):4–8.

Zalasiewicz, Jan. 2008. *The Earth after Us: What Legacy will Humans Leave in the Rocks?* Oxford: Oxford University Press.

Zelick R, and P. M. Narins. 1985. "Characterization of the Advertisement Call Oscillator in the Frog *Elutherodactylus coqui.*" *J. Com. Physiol.* A 156:223–229.

Zeuner, Friedrich E. 1963. *A History of Domesticated Animals.* London: Hutchkinson.

# Index

# About the Contributors

**Almo Farina** is professor of ecology at the Department of Basic Sciences and Foundations, the University of Urbino (Italy). He is also director of the International Institute of Ecoacoustics, and president of the International Society of Ecoacoustics (https://sites.google.com/site/ecoacoustics-society/). Farina is also a member of the Governing Board of the International Society of Code Biology.

**Carlo Brentari** is adjunct professor of Hermeneutics at the Department of Humanities of the University of Trento (Italy). In 2002 he obtained a doctorate in philosophy at the Karl-Franzens-Universität of Graz (Austria). His main research fields are the German philosophical anthropology of the twentieth century, the ethology of Konrad Lorenz, and modern and contemporary theoretical biology (in particular the work of Jakob von Uexküll). His current research is devoted to Nicolai Hartmann. Most recent publication: *Jakob von Uexküll. The Discovery of the Umwelt between Biosemiotics and Theoretical Biology*, Springer, 2015.

**Katharine Dow** is a research associate in the sociology of reproduction at the University of Cambridge. She received her PhD in social anthropology from London School of Economics, then completed an ESRC postdoctoral fellowship at the University of Edinburgh. She is currently working on a book based on her doctoral research, which concerns how people involved in the environmental movement and cetacean conservation in Scotland think about reproduction and the ethics of assisted reproductive technologies. She has a longstanding interest in people's relationships to and with nature and how ideas of naturalness are implicated in morality and ethics.

**Martin Drenthen** is associate professor of Philosophy and coordinator of the Centre for Nature and Society at the Institute for Science, Innovation and Society at Radboud University Nijmegen. He is project leader and principal investigator of the research project "Reading the landscape: A hermeneutic approach to environmental ethics," about ethical issues regarding rewilding, conservation of landscape heritage, and cultural identity. His research topics include environmental hermeneutics, landscape philosophy, ethics of place, the ethics of ecological restoration, and the impact of rewilding on cultural environmental identity.

**Annabelle Dufourcq** received her PhD in Philosophy from the University Paris I, Panthéon-Sorbonne in 2008. She has published two books: *La dimension imaginaire du réel dans la philosophie de Husserl* (Springer, Phaenomenologica, 2010) and *Merleau-Ponty: une ontologie de l'imaginaire* (Springer, Phaenomenologica, 2012). She served as the guest-editor for the eleventh issue of *Environmental Philosophy* (2013). She works at Charles University (Faculty of Humanities) and at Radboud University as an assistant professor.

**Peter Gaitsch** gained his PhD from the University of Vienna in 2013 with a thesis in philosophy (published *Eric Weils Logik der Philosophie. Eine phänomenologische Relektüre*. Karl Alber Verlag 2014). He is a faculty member at the Department of Theology, University of Graz. His current research interests are in the fields of phenomenology, philosophy of biology, and philosophy of religion.

**Gisela Kaplan** is professor at the University of New England and Honorary Professor at the Queensland Brain Institute, Australia. She has conducted groundbreaking research into vocal learning, communication, and cognition in vertebrates and become an influential voice on animal sentience. Professor Kaplan, an award-winning author, has published over 250 research articles and 21 books—her latest called *Bird Minds* (2015). She holds two PhDs (vet science/social science) and an honorary DSc for her contributions to life sciences.

**Eva Meijer** is currently working on a PhD project in philosophy at the University of Amsterdam, titled "Political Animal Voices," in which she develops a theory of political animal voice. She teaches the course "Animal Ethics and Politics" at the University of Amsterdam and is the chair of the Dutch study group for animal ethics. Recent publications include "Political Communication with Animals" in *Humanimalia: A Journal of Human-Animal Interface Studies*, and "Stray Philosophy: Dog-Human Observations on Language, Freedom and Politics" in *Journal for Critical Animal Studies*. In addition to her academic work, Meijer works as a novelist, visual artist, and singer-songwriter.

**Kristin Armstrong Oma** is an archaeologist with a PhD from the University of Southampton, and a postdoc from the University of Oslo. She is currently associate professor of archaeology at the department of cultural heritage, Museum of Archaeology, University of Stavanger. Her research bridges archaeology and the interdisciplinary field Human-Animal Studies. Her academic work on human-animal relations in prehistory has covered topics related to horses as companion species, the social practice of humans and animals in the household arena, and the development of landscape and environment in relation to farm life and husbandry practices.

**Silver Rattasepp** is a unior researcher and a PhD student in the Department of Semiotics, University of Tartu. His research interests encompass biosemiotics and zoosemiotics, particularly in the Uexküllian vein of animal knowing, as well as animal studies, science and technology studies, and speculative realism. Current research involves inquiring if and how the diverse forms of animal knowings and subjectivities can inform various philosophical issues commonly limited to humans.

**Susan M. Rustick** (PhD in Linguistics, University of Wisconsin) is a professor at Edgewood College in Madison, Wisconsin, USA, where she teaches courses in theoretical and applied linguistics, animal studies, environmental ethics, and animal ethics. Recently presented papers include "Animal Redemption" in Bonn, Germany, and "Transforming Human Identity" in Kentucky.

**Bronislaw Szerszynski** is Reader in Sociology at the Department of Sociology, Lancaster University, UK, where he also works at the Centre for the Study of Environmental Change (CSEC). His research draws on the social sciences, humanities, arts, and natural sciences in order to place contemporary changes in the relationship between humans, environment, and technology in the longer perspective of human and planetary history. He is author of *Nature, Technology and the Sacred* (2005), and co-editor of *Risk, Environment and Modernity* (1996), *Re-Ordering Nature* (2003), and *Nature Performed* (2003).

**Mateusz Tokarski** is a PhD student at the Institute for Science, Innovation, and Society at Radboud University in Nijmegen, Netherlands. He holds a Bachelor degree in Film and Theatre Studies from London Metropolitan University and a Master's degree in Cognitive Semiotics from Aarhus University. During his master's he also studied ecosemiotics at Tartu University where he became interested in issues associated with close cohabitation of same spaces by humans and animals—a subject which he investigates in his PhD where he focuses on moral conflicts around spontaneous rewilding in Europe.

**Morten Tønnessen** is associate professor of philosophy at University of Stavanger (Norway), with a PhD from University of Tartu (Estonia), 2011. He is Norwegian project leader for the Estonian-Norwegian research project "Animals In Changing Environments: Cultural Mediation and Semiotic Analysis" (EMP151), Co-Editor-in-Chief of the Springer journal *Biosemiotics*, and chair of Minding Animals Norway. Tønnessen has co-edited *Semiotics in the Wild: Essays in Honour of Kalevi Kull on the*

*Occasion of His 60th Birthday* (2012) and *The Semiotics of Animal Representations* (2014).

**Sebastjan Vörös** is an assistant professor and a Research Fellow at the Faculty of Arts (University of Ljubljana). His main topics of interest encompass philosophy of mind, phenomenology, philosophy of science, cognitive science, and philosophy of religion. He has published several articles and translations, and is the author of a book entitled *The Images of the Unimaginable: (Neuro)Science, Phenomenology, Mysticism* (in Slovene).

**Louise Westling** is professor emerita of English and Environmental Studies at the University of Oregon. Books include *Sacred Groves and Ravaged Gardens: The Fiction of Eudora Welty, Carson McCullers, and Flannery O'Connor* (1985); *The Green Breast of the New World: Landscape, Gender, and American Literature* (1996), and *The Logos of the Living World: Merleau-Ponty, Animals, and Language* (2014). She is also editor of *The Cambridge Companion to Literature and the Environment* (2014).